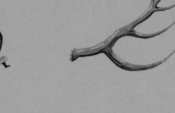

A micro-climate

WOODSTON

micro — macro.

Also by John Lewis-Stempel

Meadowland: The Private Life of an English Field

The Secret Life of the Owl

The Wildlife Garden

Foraging: The Essential Guide

Fatherhood: An Anthology

The Autobiography of the British Soldier

England: The Autobiography

The Wild Life: A Year of Living on Wild Food

Six Weeks: The Short and Gallant Life of the British Officer in the First World War

The War Behind the Wire: Life, Death and Heroism Amongst British Prisoners of War, 1914–18

The Running Hare: The Secret Life of Farmland

Where Poppies Blow: The British Soldier, Nature, the Great War

The Wood: The Life and Times of Cockshutt Wood

Still Water: The Deep Life of the Pond

The Glorious Life of the Oak

The Wild Life of the Fox

The Private Life of the Hare

WOODSTON

The Biography of an English Farm

John Lewis-Stempel

doubleday

TRANSWORLD PUBLISHERS
Penguin Random House, One Embassy Gardens,
8 Viaduct Gardens, London SW11 7BW
www.penguin.co.uk

Transworld is part of the Penguin Random House group of companies
whose addresses can be found at global.penguinrandomhouse.com

First published in Great Britain in 2021 by Doubleday
an imprint of Transworld Publishers

A CIP catalogue record for this book
is available from the British Library.

ISBN 9780857525796

Typeset in 14.25/18.5pt Granjon LT Std by Jouve (UK), Milton Keynes
Printed and bound in Great Britain by Clays Ltd, Elcograf S.p.A.

The authorized representative in the EEA is Penguin Random House Ireland,
Morrison Chambers, 32 Nassau Street, Dublin D02 YH68.

Penguin Random House is committed to a sustainable future
for our business, our readers and our planet. This book is made
from Forest Stewardship Council® certified paper.

Modern history has been much too sparing in its prose pictures of pastoral life. A great general or statesman has never lacked the love of a biographer; but the thoughts and labours of men who lived remote from cities, and silently built up an improved race of sheep or cattle, whose influence was to be felt in every market, have no adequate record.

H. H. Dixon (1869–1953)

On English ground
You understand the letter, — ere the fall,
How Adam lived in a garden. All the fields
Are tied up fast with hedges, nosegay-like;
The hills are crumpled plains, — the plains, parterres,
The trees, round, woolly, ready to be clipped;
And if you seek for any wilderness
You find, at best, a park. A nature tamed
And grown domestic like a barn-door fowl,
Which does not awe you with its claws and beak,
Nor tempt you to an eyrie too high up,
But which, in cackling, sets you thinking of
Your eggs to-morrow at breakfast, in the pause
Of finer meditation.

From *Aurora Leigh*, Elizabeth Barrett Browning (1806–61)

CONTENTS

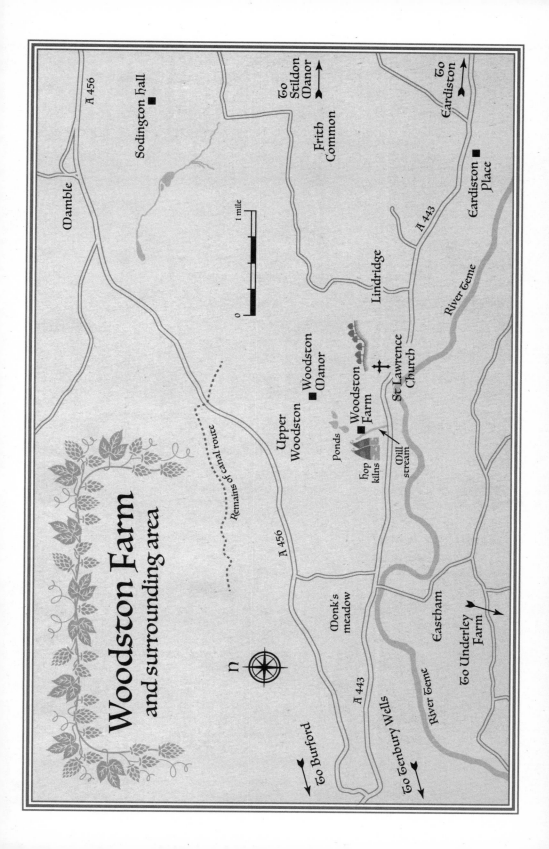

PROLOGUE

Et in Arcadia Ego

A dedication to Joe and Margaret Amos

THE GRAVEYARD IS SHROUDED in mist from the river.

I've come here to meet people I cannot see, but the mist is blameless. My people are long dead.

In the dark tower of a yew tree a wren *tisks* its alarm call, as regular as the counting of clocks. Further away, in a vague fir, a robin sings an autumn hymn, muffled and melancholic.

Then, there they are: Percival Amos and Margaret Amos, their names inscribed in black on a grey marble headstone. The roots of the fir have tilted the stone; ivy clutches at it. Clagged with damp, their memorial needs a wipe of my coat sleeve for their names to shine.

I loved my maternal grandparents, and not just because they cared for me during tranches of my childhood, but also for what they were. They were the people of Thomas Gray's 'Elegy Written in a Country Churchyard':

Oft did the harvest to their sickle yield,
Their furrow oft the stubborn glebe has broke;

1

How jocund did they drive their team afield!
How bow'd the woods beneath their sturdy stroke!

Percival – known to everyone as Joe – and Margaret Amos were yeoman farmers, salt of the earth, the backbone of England. My grandmother's ancestors fought at Agincourt as men-at-arms; they were hard people, hard like the Herefordshire mountain they kept their sheep on.

For most of their adult lives – they married in their early twenties – Joe and Marg were tenant farmers, of hops, and for a decade or more farmed up the lane from here at Lindridge.

Time flies; ivy grows. My duty this morning is the pulling away of the evergreen parasitic tendrils.

I was last here in this churchyard beside the River Teme in the long hot summer. It was midday in late June, and a young guy in a baseball cap was varnishing a bench, encouraged by his girlfriend and her radio, though this was turned down reverently; a wood pigeon was roo-cooing in the heat, soft and somnolent. It was peaceful, and it was beautiful too. Bounded by the sparkling Teme and a row of red-brick cottages, St Mary's in Tenbury Wells, Worcestershire, is the perfect English graveyard.

That day in June, with not a cloud in the sky, the sun blinded off the weathervane, a gold cross, on the Norman steeple. A sparrow flew to its nest, a hole in the stone of the nave. Swallows plastered their cup-home on an oak beam in the porch. The church gave both birds refuge.

The grass was full of flowers as I wandered down between the gravestones. Ox-eye daisy, borage, white clover, dandelion, meadow vetch.

In almost every floral-coloured aisle, an echo of a name gone now, a call on conscience and memory. Lambleys. Wilcocks. Yarnolds.

Kin.

Cabbage-white and meadow-brown butterflies blatted about in the silent air, indecisive, before settling on the lilac. There were mason bees, and the elder's fruits were green and vital.

Odd, is it not, that to see living things today one often needs to visit the place of the dead? The churchyard of St Mary's is God's Acre. Today, in the great white peace of an October morning, a single thrush starts singing matins in the yew behind the nave.

I'm not sure how old the yew trees of St Mary's graveyard are. *Taxus baccata* can live for thousands of years. But I do know they have seen generations of my family come into this life, leave it. When I see the yews they recall to mind – always – four particular lines from Gray's 'Elegy':

> *Beneath those rugged elms, that yew-tree's shade,*
> *Where heaves the turf in many a mould'ring heap,*
> *Each in his narrow cell for ever laid,*
> *The rude forefathers of the hamlet sleep.*

My rude grandparents sleep here, in their native red soil. Today, from a yew as feathered as a bird itself, a thrush sings, louder and louder in the mist in this corner of England where Herefordshire, Shropshire and Worcestershire meet.

I cannot carve stone. So, I wish to fashion a memorial in print to Joe and Margaret Amos, and all the yeoman farmers of England, past, present and future.

This is what follows.

INTRODUCTION

I remember Woodston . . .

IF GOD HAS DESIGNED a Typical English Setting, it is surely
Woodston Farm on the Herefordshire/Worcestershire/
Shropshire border. The land to the front of the farm is flat
and runs down to the River Teme, and beyond to woods;
behind and to the sides it is gently hilly, suitable for slow,
fat sheep. There are still some limes or 'linden' on the rising
land that gives the surrounding parish its name, Lindridge,
or its various corruptions over the centuries: Lindericgeas
(eleventh century); Lynderug, Linderugge (twelfth century); Lindruge, Lindrigg (thirteenth century); Lyndrugge,
Lindriche (fourteenth century); Lynderige (sixteenth century). The nearest property to Woodston is the church of St
Lawrence, reached by a footpath and a five-minute walk
to the top of the linden ridge, which begins the east end of
its sharp, sheep's-back arc in Woodston's fields. There is no
village of Lindridge as such, only the church of St Lawrence, the Georgian-faced rectory and a school; the nearest
settlement to Woodston is Eardiston, a hamlet a mile and
a half to the north. That is not to say that Woodston is
isolated – among the historic reasons for its relative success

as a farm is its proximity to transport systems, beginning with the ancient road to Tenbury Wells, the nearest town – but it lives in a small pool of its own tranquillity. The parish of Lindridge comprises 2,496 acres of land, of which 21 are inland water. Lindridge figures prominently in the life story of Woodston for a simple reason: Woodston as a distinct working farm was only named in the 1300s; before that, it was merely unnamed land in that essential unit of English physical and legal recognition, the parish.

The soil of Woodston is marl, clay and sandstone; the subsoil is Old Red Sandstone (and more of that anon, since the soil is the beginning and, maybe in these chemical days, the end of the farming story). Once, Woodston boasted some of the finest hopyards in England, watered from the gin-clear Teme (a tragically misnamed river; the name derives from the Celtic 'tamesis', meaning 'dark', 'Thames' coming from the same root). The hops have gone, the brick hop kilns converted into bijou, gleaming apartments. (Well, farmers were told to diversify.) But Woodston remains a traditional working farm, part arable, part livestock. Family owned.

I think I was seven when I first went to Woodston; we, my mother and I, walked up the long farm drive, past the orchard, to look at the farmhouse. She was on a nostalgia trip.

My mother grew up at Woodston, where my grandfather, Joe Amos, was the farm manager. Joe Amos managed Woodston for about thirty years, with a few breaks, when he stormed off in dispute with the agent of the absentee owner, the somewhat sinisterly named Farmer Pudge, a major local landowner. My grandfather, known to his grandchildren as 'Poppop', always went back. He was always welcomed back, because he was an acknowledged master grower of hops.

After managing Woodston, my grandfather managed Woodend at Cradley (part of the Duchy estate) then, semi-retired, worked as the bailiff – an antiquated expression, but essentially a farm overseer – at Withington in East Herefordshire. When he finally, *finally* retired from farming at seventy, he and my grandmother bought a house along the road from Woodston. My parents were married down the road at St Lawrence's in 1952 and had their wedding reception in the bagging room of the hop kiln at Woodston.

For many years I have wanted to write the life story of a farm, but which farm? For a while I was tempted by the 'biography' of my own sheep farm high in the Welsh Borders, but its ovine monoculture barred half the farming story, the raising of crops as well as livestock. So Woodston became the chosen place, partly through sentimental family attachment, partly because it was the archetypal English farm, mixing animal husbandry and crop growing. I could find no better example. It is England's farm.

Also, such a manner of farming is not only the story of the best of England's farming past, it is the route-map to a better farming future, more profitable, more Nature-friendly.

This, then, is the biography of Woodston, from the creation of its DNA – the very soil on which it thrives – four billion years ago to AD 1950. And 'biography' is the correct term, because farms live, have character, personality.

Few of Woodston's farmers and labourers left written records. Archaeology reveals a little more of the picture, as does folk memory (tales my grandparents and my mother told me), as does the 'lie of the land' – farmland can be interrogated by anyone with a trained eye. A field with a corner gate, for instance, is almost certain to have been founded as a meadow for livestock, since beasts are easier to herd or drive through such an arrangement (the 90 degrees where two lines of hedge or wall meet act as a funnel) in distinction to a hedge with the gate in its middle.

However, these methods together are insufficient for a complete biography of a farm. They are not enough. Accordingly, sometimes I have engaged in what I term 'Method Writing'™ – I have tried to live and work in the style of the period concerned, such as ploughing with an antler like the Prehistorics, eating medieval peasant potage. *Woodston* is tethered by factuality but not chained by it.

The English countryside was improved by our ancestors after it was created by Nature; England's landscape is the result of *agri-culture*. This is one reason why the

English landscape, particularly its 'classical' quilt pattern of fields and hedges, causes such devotion. It is shared heritage. As the poet laureate John Masefield versed during the Great War, when the patriotism of so many boys and men was fired by English pastoral scenes, the land is 'the past speaking dear'. He knew, because he lived in this self-same heart of England, at Ledbury.

The English countryside is a work of human art, done by the many and the nameless: farmers and farm labourers.

My family. Yours too.

CHAPTER I

EARTH

From the void to the making of the soil that is the
foundation of Woodston and English farming – The
volcanic eruptions, the inundations, and glaciations that
produced the current course of the Teme – The wildwood

In the beginning was the earth . . .

THIS PLACE HAS NUMBERS, as well as letters. The coordinates of Woodston are 52° 19' 2.12" N, 2° 28' 39.43" W. That is its current location on the globe, and 'place' in the life of a farm makes for three inescapable inheritances: geology, climate, angle of relationship to the sun. The waxing and waning of the moon has an influence too; the stars, unlike the navigational aid they provide those at sea, have no role, except to charm the scene.

In the beginning there was nothing here. Only void. The Ancients believed in four Elements, those of Air, Fire, Water, Earth, and while we might sneer at their science, the physical early history of Woodston is exactly a story of these things. From the nothing of the void came,

11

via the 'Big Bang' of 13.5 billion years ago, the gaseous cloud (Air), which reduced to a burning ball (Fire), then to something solid (Earth). Order out of chaos. By 600 million years ago there was *terra firma* in the place that one day would be called Woodston. After Fire, Earth, came Water. Life in its earliest forms appeared 3.8 billion years ago, complex life 570 million years ago; it slowly evolved in the warm Odovician and Silurian seas which covered Woodston then; relics of those primeval seas are the Silurian coral reefs at nearby Ludlow.

Each succeeding geological epoch laid down sedimentary rocks, each layer a stony-soily epitaph to that time. We think about history as coming down to us; but creation, generally, builds upwards, layer on top of layer. A land building upwards. After the coming of life, these geological eras had one common characteristic. They were necropoli of dead fauna and flora. The rocks and soil beneath us at Woodston are boneyards. Or, in the prosaics of farming, fertilizer. This earth, this soil-food at Woodston edges to acid, with a pH of 4.0 to 5.4, meaning it will support acid-longing plants such as sorrel, sucked by farmworkers when a-haying to slake their thirst.

Only on paper are geological sequences smooth. The land we call Britain, as it drifted around the globe, spasmed with fractures, folds and faults. Around 400 million years ago a clash of tectonic plates threw up giant mountains around Woodston. A bird flying over Woodston today can see their eroded stumps in the Malverns, Clee

Hill and, in the far dramatic distance to the west, the Black Mountains of the Welsh border.

In the Devonian period Woodston was south of the equator; it was a fiery desert dominated by prototaxites, 20-yard-high fungal spires; the red, haematite colours of this badland still glow in Woodston's soil. Thus the geology of Woodston is officially 'Old Red Sandstone Raglan Mudstone Formation (Silurian/Devonian)'.

There is something literally grounding about putting a finger into this earth. If you stick your index finger in this Woodston soil, as I did the other spring day, wriggle it around and withdraw it, you will find your finger stained with the pink-going-on-scarlet of human blood, and see why people once thought humans were moulded by the divinities from clay. The word human and the word humus, meaning 'soil', even come from the same root in Proto-Indo-European, the ancestral language of the European family. The Hebrew *adam*, meaning 'man', is from *adamah*, ground. We live off the earth, and when we die we go back to it. To add to the humus.

The picture of Woodston, as it piggybacks northward on a tectonic plate, turns tropical. Three hundred million years ago, Woodston was a fetid, insect-infested everglade straddling the equator, whose forest of giant ferns fossilized as Carboniferous rocks – coal, the thin seams of which would one day be mined at Mamble, on the edge of the Wyre Forest coalfield, ten minutes from Woodston Farm as humans walk. This coal would light

Woodston's fires for eras and epochs, and into my grandfather's heyday, the 1940s.

Over the vastness of biological time – but the blinking
of an eye for a Creator – the strange creatures that crawled
from the Devonian ooze become stranger still. The first
insects – the farmer's salvation for their pollinating, but
the farmer's curse for their destruction of crop and beast –
shuffled around in the ferns of Woodston around 479
million years BC. In the Mesozoic era, giant reptiles
haunted Woodston's land, and darkened the sky overhead with their wings.

It was a time of original noise: the jaw-grinding of
herbivores grazing, the scream of animals losing their life
as the talons of archaeopteryx sank in. These were proto-
farmland sounds, their echoes heard millennia later in
the cows eating down in the river meadow, the cry of the
chicken taken by the hawk.

During the final 60 million years of Mesozoic time,
England subsided, leading to deep inroads of the Creta-
ceous seas, the tropically warm waters of which teemed
with tiny marine organisms whose skeletons dropped in
an endless white blizzard of death. Where they fell they
buried other corpses, including shoals of primitive
sponges which, over time, became the flint knobs from
which early humans would fashion tools to cut down the
wildwood.

Only with the 'Alpine Storm', a 20-million-year-
long global shiver, was England raised from the waves.

By now the giant reptiles were gone, along with the primitive tree-ferns and giant horsetails of the coal forests. In their place came plants and animals more closely related to the flora and fauna of today. Sabre-toothed tigers and other fabulous beasts – elephant, elk, cave bear, woolly rhinoceros, bison – roamed the Woodston wilderness.

In western Scotland, volcanoes still poured lava, but Woodston was becoming docile. Never again would it be a place of extremes. Henceforth it would always respond to gentler rhythms. The two or three million years of the Pleistocene period brought the Great Ice Age (water, again), a frozen, smothering white tide that slid down Britain, groaning and grinding away at the land, throwing up great crystalline, sparkling cliffs.

Woodston escaped the glacial sheet, though the ice-cold left its mark on the landscape. The gentle hummocks behind and opposite Woodston today are the result of solifluction, where the upper layers of permafrost melted in summer, allowing the top six feet or so of ground to slide down over the more frozen base.

Meltwater drained into the Teme valley from the west, creating a lake. Under Woodston the groundwater still follows the routes of this glacial era, but the surface at some indeterminate moment in the past began draining to the east, a mismatch causing the long record of flooding around the farm in our historical times. In a submission made to Parliament during Cromwell's Protectorate, the inhabitants of Knighton, two miles away, declared that Lindridge was too far to travel to church, because 'the ways thereof [are] very fowle and deepe in time of winter'.[1]

More importantly for the life of Woodston, the presence of ice ten miles away at Woofferton caused the River Teme to reverse direction and find a new passage eastwards from 12,000 BC. This new river course would later provide the southern boundary of Lindridge and, eventually, of Woodston Farm itself.

When the ice retreated for the last time, in about 8500 BC, and the glacial lake at Woodston disappeared, leaving in its wake meres and mires, there came the two thousand years of warm, dry weather known as the Boreal period. The bitterly cold tundra was slowly repopulated by trees, and a fertile earth recovered. Dwarf birch was first to recolonize; then stands of true birch, the ice queen of trees, then juniper, willow, aspen and Scots pines, hazel, alder, oak, followed by elm and lime, including the rarer small-leaved lime. The latter still lives on in the Teme valley, particularly in the dingle streams that go down to the river.

~

A word about lime.

'Lime' is an altered form of Middle English *lind*, cognate to Latin *lentus* ('flexible'), fair enough for a tree which, when young, is elegant-limbed, skin-smooth. The English

We walk so minuscule in such a great story.

'lithe' is from the same root. The seventeenth-century arborist John Evelyn noted regarding lime wood:

> And because of its colour, and easy working, and that it is not subject to split, architects make with it models for their designed buildings; and the carvers in wood, not only for small figures, but large statues and intire histories, in bass, and high relieve; witness (besides several more) the lapidation of St Stephen, with the structures and elevations about it: The trophies, festoons, frutages, encarpa, and other sculptures in the frontoons, freezes, capitals, pedestals, and other ornaments and decorations, (of admirable invention and performance) to be seen about the choir of St Paul's and other churches; royal palaces, and noble houses in city and country.

The lime tree is rich in lore. In the age of chivalry, the 'Lind' was the tree of love; in the Second World War, its wood provided the frames for Mosquito aircraft. The yellow flowers are the flavouring for 'lime cordial'. The lime features in the poem 'The Owl and the Nightingale', *circa* AD 1200, the earliest mention of a hedgerow tree in English literature. Perhaps, most appositely, the ur-botanist Carl Linnaeus owed his surname to the linden tree outside the family home.

Such is the interest of lime to the artist, the lover, the warrior, the scientist, the cook, the arborist. For the farmer, large-leaved lime *Tilia platyphyllos* and small-leaved lime *Tilia cordata* are shade for livestock; they can be coppiced for fuel, bean-sticks, and – usefully for Woodston – hop-poles. The heart-shaped leaves are fodder for cattle, sheep, goats, while the fibrous layer of under-bark called 'bast' can be twisted into ropes or used to make sandals, even cloth, and was so woven by Neolithic people. The flowers are pollinated by honeybees and other insects visiting for their nectar. The wood of lime burns bright and white, fast but hot. I know because we have lime trees in our garden, pollarded every several years, the limbs unwanted for poles going on to the sitting-room fire.

Wood was the first serious fuel. We humans have been utilizing the wildwood for warmth for five hundred thousand years. Changes in climate over the last millennia have caused lime to diminish, but once it crowded the elephants and rhinos of lowland England. Touch a lime tree in a wood and you touch the ancient past.

The forest of the post-Ice Age flourished, and from shining sea to sea, England was covered by a canopy of trees. G.M. Trevelyan, in lines I learned at school, conjured the English scene then as one of innumerable treetops which 'shivered in the breezes of summer dawn and broke into wild music of millions upon millions of wakening birds: the concert was prolonged from bough to bough with scarcely a break for hundreds of miles over hill and plain and mountain'. This fantastic dawn chorus was unheard by man save where, at rarest intervals, a troop of skin-clad hunters, stone-axe in hand, moved furtively over the woodland floor.

What did the first Neolithics see when they scrambled up the salmon-pink bank of the Teme at Woodston? A wooded scene, but also a place of kind climate, a green land. Put a Neolithic man or woman at Woodston today, and he or she would in some way know it still: a place where it rains enough, but not too much, and there is ample water in summer from the river. Where the climate is mild. Where the land is sheltered from the whipping winter winds by comforting low hills. Where the soil is good.

The Neolithics came to Woodston around 3000 BC,[2] when Britain was already cut off from the Continent. They began making holes in the limitless canopy of the wildwood. They began the farming of England.

CHAPTER II

THE CLEARING

*The beginning of farming in the Woodston wildwood by the
Neolithic people – The coming of domestic sheep and
cattle – Field patterns established – The continuation of
hunter-gathering – The making of the first English hedges –
Tree hay – By the late Iron Age (Christian Year Zero) half
the wildwood around Woodston cleared for farming – The
unacknowledged plenty of prehistoric farming*

AT THE TAIL END of the 1700s a Neolithic polished stone
tool was found in Lindridge parish, a lucky find, given
the sparsity of Neolithic artefacts in Worcestershire, a
county with no landscape remains of the Stone people.
The nearest standing-stone witness to the presence of the
Neolithics is the henge circle at Mitchell's Fold, just north
of Bishop's Castle, Shropshire; the nearest long barrow is
at Dorstone in Herefordshire's Golden Valley, under the
shadowy wall of Mid-Wales.

The Lindridge stone tool was troved in a toothill at
Dodenhill (half a mile from Woodston); the toothill, a

man-made lookout post, was in the process of being flattened by workmen, the stone sought for other uses, in the endless, eternal recycling of farmland materials. Six feet down, in a bed of gravel, was the Neolithic tool, finely polished, 4¾ inches long, 1 inch wide and ¼ inch thick. At one end a sharpened cutting edge; at the other three holes drilled into the stone, presumably for a handle. The object was acquired by a local antiquary, the vicar of nearby Great Witley. It was an age when parsons did such things.

Stone suitable for making tools does not occur naturally near Woodston, or indeed most other places in England. The right stuff has to be hard but workable, capable of keeping a sharp edge. So, stone tools were mined in quarries with antlers (the Neolithic flint-miners of Grime's Grave in Norfolk used over 50,000 antlers as picks), then traded across Britain, even to and from the Continent. A well-known trade route, the Clun–Clee Trackway, a snaking earth-path running just north of Lindridge, operated in Neolithic times. It started at the Kerry Hills and led to the Severn at Bewdley. Another, lesser trackway – a branch-line – ran parallel to the Teme, and is today the A456 Tenbury to Newnham Bridge road, lying between the farmhouse and the river; Woodston was thus between a duo of trade routes, two of the hundreds that spiderwebbed Neolithic Britain. In the ensuing centuries these self-same Neolithic trade routes would be rescribed on the landscape around Woodston by canals, tarmac roads and railways. The landscape is a palimpsest, on which each epoch etches its lines of communication, of trade.

The geology of the Lindridge axe-tool suggests it is composed of the same Welsh stone from the Preseli hills transported to Wiltshire to construct Stonehenge. Other Stone-time tools have also been found in this north Worcestershire parish: a large hand axe and a loom weight; across the Teme at Eastham, a collection of flint implements was discovered in 1936 during excavation for the New Road at Hanley William. (They are exhibited in Eastham church, a matter of local pride; my mother took me to see them when I was nine. For weeks after, I tried to make fires by striking flint rocks.)

The New Stone Age people at Woodston lived a double life, part nomadic hunter-gatherers, part tyro-farmers. Foraging and farming are too easily counter-poised, too frequently regarded as mutually exclusive. Our precursors were less adamant about what was weed, what was crop. Tollund Man, of Danish prehistory, had porridge containing cultivated barley and feral fat hen (*Chenopodium album,* also known as wild goosefoot) for his last breakfast before he was hanged. The wildwood glades where the prehistoric people grazed their cattle and sheep were the same glades where they hunted red deer, wild swine, wild cattle. A glade, as well as providing space for crops, offered a clear line of sight for a flung spear, an arrow from a bow.

They hunted in earnest, because meat was good and filling: the decline of the big beasts of England, accordingly, was steep. The bears and aurochs were gone from Woodston within a millennium; of the bestiary of large mammals, only red deer survived beyond the Middle Ages locally. The animal extinctions were tragic, though we should not lightly despise the bravery of the prehistoric farmfolk who, with a pathetic spear or arrow, sought to bring down a wild bull, horns a yard wide, and which was as quick witted as it was quick at turning on its hooves. The *toros* of Pamplona and Spanish bullrings are sad, plump facsimiles of the rampaging cattle of yore.

Down to my childhood in the 1970s, farmfolk routinely picked hedgerow fruits for jam, game-hunted over the land, fished the pond and the stream. The land was expected to provide in all ways, the tame and the wild. It seemed natural. I made my own children pick nettles for soup, sorrel for sauce as late as 2010, but even at the time I was aware it was a constructed middle-class experience, one they had to be persuaded into. Foraging was not a given for them, or a necessity, as it was for my mother, my grandfather, or even for me on Broadmoor Common, above Woolhope, when no blackberries picked equalled no jam made in my stepmother's iron cauldron on the cooker from the Electricity Board showroom.

The Neolithics' antique pattern of farming and foraging died in 1996. I saw its end, on a hillside in the Golden Valley, above the medieval abbey of Dore, when an old

24

farming couple, Eric and Molly, drove 25 miles in their Land Rover 110 to pick up the windfalls from the crab apples, so 'they would not be wasted'. Nobody would bother now, and in the west of Europe the cessation of farmfolk-foraging was general.

I spent last autumn in France, where walnuts grow by the laneside, and I was the only person in the village who picked them.

So much for foraging. Neolithic *farming* required the creation of glades of pasture for the semi-domesticated herbivores, glades of bare earth for the crops. For these open spaces, the Neolithic peoples needed to clear land of trees, and for that task they needed tools of stone, such as the Lindridge axe-tool.

The Stone-Agers' prime method of clearing the trees was crudely effective. They ring-barked selected trees by chipping with a stone-axe a weeping mortal skin-wound, in imitation of the deer. Pollen records suggest a decline between 4000 and 3000 BC in tree species such as ash and an increase in plants such as nettles and plantains. As ash tends to grow on the edges of woods, its decline is associated with the clearance of tranches of the wildwood for agriculture. Stone was not their sole tree-razing tool. In the unlikely happenstance of a long, hot English summer, the Neolithic clan at Woodston may have used fire (the ash from which would have also acted as fertilizer), although English trees do not burn easily, and felled trees need to be cut up for firewood. The Neolithics also employed their livestock in clearing the land, and keep-ing it that way.

Certainly, there is no archaeological evidence for Neo-lithic use of livestock for razing the wildwood. But it stands to farming reason. In the twenty-first century I've renovated woodland choked by bramble (a domestic invasive species) and sycamore (an alien invasive species) by sending in the livestock – cows to blunderingly clear paths, trample the understorey, and then sheep to browse, pigs to rootle.

The grazing of the Neolithics' cattle, sheep and pigs, by their gorging of emergent tree shoots, stymied re-generation by the wildwood on the linden ridge at Woodston, as elsewhere. Primitive sheep would also have stripped many species of tree of their bark, a sweet favourite food. Some years ago, I put ten Shetland sheep on a borrowed paddock by the Dulas brook, in the peace of the Black Mountains, on a balmy March afternoon; next morning, the green grass was untouched, but every hazel in field and hedge was de-skinned from ground to four feet high, as far up as the sheep could reach stand-ing on their thin back legs. The white, suppurating wounds of the hazels were accusing in the dawn light. (The owner of the paddock, Brian Edwards, was very understanding.)

I have reasoned this conclusion concerning the use of livestock in clearing trees and scrub: the Neolithics would simply have noticed, noted, the ways and effects of their herbivores, and understood that all they themselves were required to do was work with the animals' natural appetites. A wood to which sheep or cattle are often admitted loses its woodland plants and quickly turns to grassland with trees.[1]

The Stone-Agers were not separate from Nature. They were of it, but beginning to take a human stance, which is to control Nature. Which is farming.

~

If a primitive farm animal such as the Soay sheep was to be utilized as a portable live brush-cutter, a warm-blooded ground-clearance instrument, it would need to be fenced in. Otherwise it would roam, 'cherry-picking' rather than nihilistically razing.

Which is why this January morning I am in a Herefordshire copse near Woodston, constructing a five-foot-high 'dry hedge' of boughs and sticks. I am putting my personal theory into my personal practice.

Since there was hard frost last night the earth is only just thawing out. (A daily Ice Age and Aftermath Melt; Nature is all cycles within cycles.) From the bare trees around me, the meltwater drips as wet as tears. Further out in the wood, a jay screeches, which is the coldest sound of the English winter. Two robins up in opposing hazel bushes 25 yards apart sing their territorial songs; building their own sort of fence, invisible in the sky.

The oddity of manual work is that it encourages

philosophy, as per the venerable Chinese aphorism, 'If you want wisdom, ask the fellows who cut the hay.' Thoughts keep rising like bubbles, as I pound sharpened branches into the ground with a round stone, putting the driest, most brittle stuff aside for a hearth fire; the Stone Age could equally have been termed the Wood Age. At one point I stop to de-pith some glistening red rosehips for a snack. The sheer scale of such work, days and days, must have engendered a radically different sense of time. When you have for ever, the only time that counts is now.

The Stone-Agers would doubtless have undertaken a task such as fence-making cooperatively, as a clan. My sole companion in our supposedly progressed Modern Times is a black Labrador puppy, Plum.

I start to flag, and decide to spend ten minutes on dog-training instead. Of course, if you think about it, teaching a puppy is only a replication of our original taming of the wolf. If you think about it more, the wolf-dog, which is menacing to sheep, can be used as a furred, running hedge to contain and control sheep. A sheepdog, in other words.

Herding. Men and boys did it too, carrying sticks. (And still do; I have a long shepherd's crook, which can be waved like a giant hand behind the flock to guide them; held right, they go left; held left, they go right. My stick for the cows, the 'cush stick', is a short hazel pole cut from a bush in the hedge by the house. Some of us still 'make do'.) Up in twenty-first-century Cumbria, much is made of 'hefted' sheep, ovines trained to remain within definite borders, the knowledge of the cordon passed down from ewe to lamb.[2] Perhaps hefting was employed in misty, ancient times elsewhere in these isles?

Whatever. The 'dry hedge' worked to contain our Hebridean sheep, which are as close to the primitive Soays as you can get. They annihilated the 25 by 25 yards of scrub to the ground within three days.

While I was constructing the dry hedge, a fox trotted past. The fox is perhaps our most ancient landowner. The species has lived in these isles since the geological Wolstonian period, 352,000 to 130,000 years ago. Time is never linear, it contorts and spasms; it takes a fox from millennia gone by to accompany my scene in AD 2020.

I saw the fox. So did my Woodstoni ancestor.

On this same day, I heard a deer barking in the wood, and some atavistic impulse tempted me – fleetingly – to pick up a shaft to use as a spear. You can take a human out of the prehistoric wildwood, but you can never quite take the prehistoric out of the human. I was confirmed in this by the family BBQ at night, bits of rare-breed beef (and vegetarian option) cooking over open smoky fire, and me glowing with the heat and the contentment, as sparks joined the stars; once upon a time all food was 'slow food'. In those night moments, my clan around me, I was Stone Age and Primary.

~

The day after, I made nettle pudding, the oldest-known dish of the British. Researchers from the University of Wales Institute have found traces of nettle pudding 8,000 years old; Neolithic chefs blended the nettles with barley flour, other greens, salt and water to make a big dumpling to go into stews.

God, that pudding was good, and it was filling, for stomach and for mind (with its green marbled beauty): firm but rubbery fawn on the outside, on the inside green marbling, a gorgeous dumpling which satisfied my theory of the long coexistence of foraging with farming, the nettles being wild and the barley being cultivated. It reminded me, too, of the importance of Fire, because without the domestication of Fire, and its ability to tenderize the previously unpalatable (cereal grains, for example), farming would have been pointless, and Sapiens would have required intestines so long that little energy would have been left over for brain work.

The Neolithics disrelished nothing that came usefully to a sweep of the sickle, a pluck of the fingers, as analysis of carbonized cakes at Glastonbury Lake village shows. They were found to be made not only from flour of wheat and barley but also that of roughly ground seeds of wild oats, brome (*Bromus secalinus*) and orache (*Atriplex pautula*). The prehistoric grain crop was weedy, and the strong flavours and coarse textures of some 'weed' seeds were accepted, even enjoyed. Of course, one should not assume that plants such as orache were regarded by the Prehistorics as wild; they may well have cultivated them.

For millennia the line between weed and cultivated crop, what was wild and what farmed, was blurred. Doubtless this led to variety in diet and was good for the guts, or the 'microbiome', in contemporary parlance. The best of both worlds.

~

Unfortunately, humans seem doomed to learn and forget. It is not just the wheel that gets reinvented, but food and medicine. The nettles of nettle pudding remained a British health tonic until the nineteenth century – hawkers on Victorian streets used to cry, 'Nettles with tender shoots to cleanse the blood', Scottish miners put them in their 'parritch' – and then in the West we suffered a collective amnesia until the 1970s, when we again grasped the nettle as a health booster, in the form of nettle tea.

Our greatest amnesia, though, has been antibiotics. In 2017, researchers discovered that a Neanderthal with a dental abscess had been eating a penicillin-producing mould, implying knowledge of its antibiotic properties. Alexander Fleming received the Nobel prize for 'discovering' what supposedly stupid Neanderthals knew millennia ago. There are other less ancient examples of medical knowledge, including Ötzi the Iceman, a Neolithic corpse found in glacial ice in the Italian Alps, dating from around 3000 BC. On the day he died, the

Iceman was carrying a pouch stuffed with wads of the birch polypore mushroom (*Fomitopsis betulina*), the curative properties of which are confirmed by modern pharmacological studies. The mushroom is antibacterial, antiparasitic, antiviral, anti-inflammatory, anticancer, neuroprotective and immunomodulating.[3]

~

A comment on hedges. Prehistoric farmers patented the 'mixed farming' that would be Woodston's story, and the best of English agriculture, part arable, part livestock, symbiotic because the one system fed the other: animal dung manured the cereal land, the cereals provendered the animals. The flaw in mixed farming was that the livestock needed to be kept apart from the cereal crop. Fences keep livestock out, as well as in.

Hence the hedge, if only a 'dead hedge' of thorn bushes or cleft stakes. (Moccas Park in Herefordshire has a medieval-type fence of split oak, eloquently able to keep deer, the boundingest of creatures, in their proper place.) But the advantages and construction of a living 'quick hedge' were already understood before the birth of Christ. To quote from Caesar's *Gallic War*:

> The Nervii [a tribe living on the borders of France and Belgium], in order that they might more easily obstruct their neighbours' cavalry who came down on them for plunder, half cut young trees, bent them

over and interwove branches among them, and with
brambles and young thorns growing up, these hedges
present a barrier which is almost impenetrable.

What was good for the Nervii was good for the Woodstoni. There is no mark of prehistoric hedges at Woodston,
but hedges are wood, and wood rots. The history of the
English hedge is eventually caught on record; in Grim-
ley, Worcestershire, the annals say, in AD 966 a 'hegestowe'
was destroyed. It was already 'ealdon' (old).

Worcestershire would always be a hedged county,
romantically English, full of blossom in spring, somewhere in summer to take shade, somewhere in autumn
to pick fruits, in winter a bulwark against bitter winds.
And always the great screen dividers of the landscape,
offering privacy when there was little in the cramped
communal construction called 'house', as well as keeping
the beasts confined in their assigned place, not wandering
far afield. The thorny 'quick hedge' – so usually associated with the Georgian Enclosures – was there at the
beginning of English farming, and is an old, old ingredi-
ent in our farming story; the great 'ripping out' of hedges
in the 1960s and 1970s was always harm against history as
well as against ecology.

Hedges, additionally, are property markers. The initial
boundary between prehistoric farmsteads was a wooden
wall: trees, belts of them. But as more land was claimed,
the trees thinned, and new markers were needed – such as

the quasi wood, the linear wood, which is the hedge. Invariably with a ditch before it, the ditch itself being a marker, as well as a drainage system. It took about five years in Stone Time to grow a decent hedge from hardwood cuttings; it takes about five years in Computer Chip Time to grow a decent hedge from hardwood cuttings. The rhythms of botany are eternal, transhistorical.

Where there was no hedge to indicate ownership, boulders were used. Whether hedge or boulder, it made little difference to the sense of property, to territory. Farmers then possessed the landscape in their minds. Every bush, rut, boulder meant to them – whether as boundary, site of devotion, place of death – the abode of the spirits.

~

Wheresoever they went, the archaeological books concur, the Neolithic people chose the ground on which it was easiest to fell trees, meaning the hilltops, where tree cover is lightest, and gravity on their side; it is easier to drag a fallen tree downwards than across or up. But archaeologists are not farmers, so there is more to the Prehistorics' choice of hilltop for their fortified farmstead than arboreal density and Newtonian physics. Psychologically and practically, high ground is easier to defend, from rustlers, from all the predators of the lurking, wolfy wildwood. Also, there is the oldest truth in livestock farming, and I saw it first when I was a toddler staying with my uncle Willi, a Gower sheep-farmer, as we watched Shadow, his collie, round up bleating, protesting ewes, these as fluffy white as the breaking waves in the background.

For years, in the most protected corner of my heart, I cherished holding Willi's weather-leathered hand, because it connected me to centuries of Man and Land, Man and Dog. Only when I started keeping sheep myself did I rerun the scene to understand the mechanics: Willi and Shadow drove the sheep *up* the salt-blasted castle mound to pen them; the natural tendency of livestock when requiring rest or respite is to head to high ground, for the lookout view and for the dryness. Safety and comfort in one place. The ancient farmers at Woodston, in needing to bring their livestock in at night for protection from wild animals, would have been cooperating with ovine and bovine inclination if home was on high.

Last week I was standing up on the linden ridge, on the steep steps of St Lawrence's, enjoying the exhilarating view. On a clear day from St Lawrence's you can see, across the Teme (vertiginously below one's feet), all the way to Wales. And it occurred to me, perhaps the Neolithics also liked a view, for the sheer pleasure of it.

The Woodstoni made the first permanent farmhouse at Woodston, on the eastern curve of the linden hill, out of the west wind, and below the high grove where they talked to the spirits. In their round house with woven hazel wattle walls, turf roof, they assembled all the clutter of permanent agricultural residence – pottery containers, quernstones, axes and sickles – and outside, in anticipation of the 'farmyard', they dug storage pits, and havened their domesticated beasts at night behind a stockade, safe from savage animals and the occasional raiders.

So, the Neolithic farmers at Woodston started their deforesting up on the linden ridge and the long north-west hummock, which lies attached like a dog's pawing front leg, creating small, irregular but essentially square-ish fields, which spread out from the farmstead in an archipelago of open clearings in the domineering wild-wood sea. (The ghost lines of these garden-sized fields, if one squints at the land around the church, materialize briefly in Close Orchard, Little Orchard, Upper Coney Green, and then drown under the agricultural workings of later times.)

The beginnings of the pastoral English landscape were composed in the Neolithic period. Those who lament that contemporary Britain has less tree cover than countries in Continental Europe miss the point; 'twas (almost) ever thus. We are agriculturalists.

~

The Neolithics' collection of farm animals in England was dog, sheep, cow, pig, in that order.

Sheep were the first 'meat animal' in the world to be domesticated. (*Homo sapiens sapiens* is, naturally, an omnivore going on carnivore, hence the 'canine teeth'.) The wild mouflon sheep (*Ovis gmelini*) was declining in numbers; domesticating them was a clever way of managing a dwindling resource during the archaeological period 9000 to 8000 BC in the Fertile Crescent of the Middle East, as likely as not in Mesopotamia, modern-day Iraq.

Sheep were imported by the Neolithics to our isles *circa* 4000 BC. Indeed the wool and hide of the sheep's back possibly allowed expansion of *Homo sapiens sapiens* to the cold north. These cautious ovines – small, dark, horned, lean like the mouflon they were developed from – bequeathed milk, wool and meat; they were not shorn, rather plucked in later summer, the fleece coming off naturally, when the yellow lanolin had crusted on their skin.

Sheep were to the Woodstoni the equivalent of the buffalo to the Sioux: the multipurpose beast. We should not think of these domesticated beasts as being akin to the over-improved, commercial livestock of modern intensive farming, their food served up to them in the form of cereal concentrate and silage (pickled grass). The Woodstoni's beasts were half wild, fending for themselves on natural grasses, roots, shrubs and trees.

Their ilk today can be seen in the Soay which runs feral on the island of that name in the St Kilda group, and any place with people with the optimism to farm them. I once spent a September afternoon in sweating sun, a dog saunter from Woodston, trying to gather a flock of Soays; I might as well have sought to gather the north wind. Only the utmost vigilance, fierce fencing and maybe dogs can have kept such flighty beasts out of the mouths of wolves. A farmer's moans of the impossibility of sheep are millennia old. They are the tales my grandfather told me. The ones I tell my children. *Plus ça change*, down on the English farm.

A farm is more than a geographical entity; it is the crucible of memory to those who ever owned it, tenanted it, worked it.

~

It is unlikely that the Neolithics of Woodston domesticated the native wild beeve, the auroch (*Bos primigenius*), to be their cow. A millennium later Julius Caesar, in his *Gallic Wars,* described the auroch as 'a little below the elephant in size, and of the appearance, colour, and shape of a bull. Their strength and speed are extraordinary; they spare neither man nor wild beast which they have espied.' These aurochs had existed since the Pliocene, and as the last Ice Age had retreated, they had proliferated across the whole European landmass into Asia and Siberia. The last auroch died in Jaktorow wood in Poland in 1627.[4]

European domestication of Ermintrude did not rely on the auroch. That beef was untameable. All farmyard cattle, according to DNA analysis, are descended from as few as eighty animals domesticated from the wild in the Zyglos mountains of Iran 10,500 years ago. The Neolithics came to Woodston accompanied by their cows, stowed in wickercraft akin to coracles and log boats, or driven bellowing through the shallow surrounding seas. Ancient cattle were smaller ('buckle height', a metre high), because reduced scale helped them survive hard times and made them easier to handle. Effectively, they were genetically engineered from the runts, bred down, not up, and with diminished horns. The preponderant race of prehistoric cattle was *Bos longifrons/brachyceros*, resembling the Kerry cow or Irish shorthorn. Farmers have always sought to select, manipulate, develop. Play little God. (And yet, Nature seeks to survive, to profligate, to continue despite our great designs. At some point aurochs miscegenated with domestic cattle, so their DNA persists today in, inter alia, the Welsh Black and the Dexter. The wild is with us still.)

Domestic imported beeves made an immediate impact on the earth of Albion; the dung beetle population increased exponentially, meaning cows early began their work as walking muck-spreaders, Nature-helpers.

~

It is easy on the archaeologist's page, the writing of 'the Neolithics cleared great areas of wildwood'. In the primitive reality of BC time before AD hydraulic machinery, it must have been backbreaking work. To kill a tree is a fraction of the land-clearance task; the unwanted tree, after being felled, needed to have its stump dug out of the ground with the scapula of beast or stick of tree. Some savvy Neolithic farmer, however, had the notion to put the dumb beasts to man's work; cattle bones in Neolithic sites show the stress-induced damage that comes from hauling. A cow, of course, is also an 'ox'.

An ox provides tractive power; as does a tractor.

There is the clearing of land, then the ploughing, the harrowing, the making of the fine 'tilth' necessary if one is to plant crops. Again, all jobs easier to describe than to do. Seed does not just drop on the surface of the ground and grow; it requires a thin cover of soil to give it darkness and moisture to germinate. In tree glades, this thin covering is given by the vegetation. The Neolithics learned that the more the soil was pulverized, the better the germination and crop quality.

On a fine English morning, on our farm in Herefordshire, I decide to plough Neolithic, with an antler, this for some reason kept by my late father when, ill, he retired to a baronial bungalow. An antler, archaeologists aver, was 'the first plough'. There are three experimental 15 by 15-yard plots before me. It is archaeology by experiment, because, frankly, I am increasingly certain archaeologists are clueless about working land, and farmers are not.

It is like the first morning of Creation, a skylark up above the wheat, the dew thick and glistening, the sky overhead baby blue. I feel I can see into Heaven and the doings of the saints.

After about twenty minutes of experimentation on this 'Day I' I am praying for their help. Plot 1 is thick sward, such as you might get in an airy wood, with few trees, where auroch have made grazing places. To get through the grass tangle to clear earth requires hours of scraping and hand pulling of weeds before the surface is scratched. A criss-cross style of scraping comes naturally. To 'harrow', to further break up the soil and destroy weeds, I use an elm branch with thick twigs (a tool, according to Celtic references, used in historic times in the west). Time taken to make an acceptable seed bed about half an inch deep? Twelve spine-stressing hours.

Day II. Plot 2 has been cleared by fire (I used a wholly incongruous paraffin flame-thrower), but the consequence of flame is to transform the surface of the soil into impenetrable ceramic. The furious heat has also killed all helpful worm activity. After six hours of swearing, I have created with hands blistered – despite gloves – a seed bed. The ash, however, dug in, acts as fertilizer.

Ah, but plot 3. Here I let the pigs go first; Old Lavender, a rare-breed Welsh pig, snout-ploughs, not only turning over the soil but digging up the roots of the most injurious weeds. With three hours of additional antler work, the seed bed is fine and ready.

Pig bones at Neolithic sites are dug up in profusion, but whether from slaughtered domesticated swine or hunted wild ones is controversial. Any swine domesticated would have been a hairy, razor-backed scavenger in the woodland, and rather 'boarish'; the British improvement of pigs did not begin until the 1770s with the

41

importation of Chinese pigs. The great benefit of pigs for the Woodstoni was their indiscriminate fecundity. Sheep and cattle breed once a year but pigs farrow twice, with great numbers of young.

I have drawn my conclusions: a Neolithic farmer, who is a mini prolepsis me (being shorter, stouter, darker-skinned), would have drawn the same. The easiest ploughing is either to contain the pigs by fencing, super-vision (by the 'swineherd') or bribery in the form of a dumped favourite food, such as acorns, so they rootle in a particular place. Or for the human to plough where the pigs have already rootled. To follow the pigs.

~

Whatever method the Neolithics employed for their ploughing, they had one great advantage over the farmer of today. The virgin soil underlying the trees, fertilized by centuries of accumulated leaf litter, was a dream growth medium. It was also replete with Nature's own plough, the earthworm. With no proper understanding of fertilizer they let fields lie fallow when the goodness of the soil diminished; their primitive system of rotation was to clear patches of forest, allow scrub to grow when soil showed signs of exhaustion and then burn off the brushwood, the wood ash acting as fertilizer. Or to allow stock to graze the small plots, preventing estab-lishment of woodland but also refreshing the worn-out soil with their droppings. The system required a lot of land – Tacitus has this to say about the Germans of his day, the first century AD: 'They change their plough-lands yearly, and still there is ground to spare.' These Prehistorics may have changed their plough-land, but they were themselves sedentary. They were rooted in their own land, not transient travellers.

～

What did the Neolithics grow on the earth of Woodston? Einkorn wheat and barley were both extensively cultivated, both imports from Asia and Africa; Einkorn wheat in its domesticated form (*Triticum monococcum*), as well as being hardy, grew vigorously enough to outpace weeds. Oats were also grown, and these were plausibly home-developed from the wild oats of the indigenous catalogue of flora. (Rye was a late and likely Saxon introduction.) The Neolithics brought with them many of our arable flowers – 'weeds', if they are in the wrong place – notably the beautiful and bright poppy and corn-cockle, which would prettify summers until AD 1950, when the chemical flood commenced. The Woodstoni harvested their cereals with flint blades wedged into a short stick, the shiny sun-catching flints brought by trader men from the raven-black land where the sun died and visible on a clear day from the linden ridge. The price of the flints being a bartered beast, a bartered exquisite pot, a bartered anything.

The more they grew into the landscape, the more

agricultural the Neolithics became, which is the essential narrative of all invaders of these isles. One of the Neolithics' developments was doubtless a primitive 'breast plough', consisting of a stout pole with a pointed end pushed by the chest through the ground. (The similar caschrom was used on crofts in western Scotland into the twentieth century.) A variant involved one person pushing, another pulling on a rope, often a man-and-wife team; English farming was a family affair. The next development in ploughing was, the aggregate of archaeologists of agriculture pontificate, the beam ard, a simple wooden plough pulled by oxen. The tilth is produced by the flow pattern of the soil around the heel of the ard.

Farmers always like a bit of kit, so why suppose the Neolithics were different? I propose that the Neolithics had more than one type of plough, and used a 'crook ard' for the drawing of seed drills.[5]

~

But cutting the earth with antler or wood would never be enough, even if hauled by cattle; the earth needed unnatural cold metal to properly cleave it.

The creation of Earth has its geological periods, the creation of the farmed landscape its technological ages. After the Age of Stone, there were, incremental in their hardening of metal, the Age of Copper (c. 2200 BC), Bronze (c. 2000–800 BC), and then the Iron Age, which came to Woodston around 700 BC. Of the ages of prehistoric farming, it is Iron that is key, with its innovations, such as the iron axe, all the better to fell the wildwood with, the metal sickle, all the better to cut corn with, the iron-tipped plough, all the better to cut the earth with. Climate change was a prime catalyst in the technological change down on the prehistoric farm; the climate of northern Europe deteriorated during the Bronze Age, *circa* 1400 BC, from dry and warm to what it is today. Damp and relatively sunless.

Sodden soil under grey and English sky required the ploughman to cut it into slices that could be turned up, to the sun, to the air. Thus the introduction of the coulter, the pointed stick or fore-share, which has no function other than to facilitate the cutting of such furrow slices. By the late Iron Age the coulter was ferrous, or at least ferrous-sheathed, meaning the Woodstoni could not only cope with climate change, but could embark on the ploughing of the heavy, wet land down towards the engorged Teme, and also towards the marshy land east-wards, a bare 750 yards from the foot of the linden ridge, run through by a stream, eventually to be canalized into a millrun and fishponds in the eighteenth century, if not before.

This Time of Iron saw an intensive campaign of woodland clearance, including the razing of the redoubt-able riparian alder and willow woods. By the end of the Iron Age, 300 BC, a swathe of Woodston had ceased to be wildwood: most of the linden ridge, and the long hummock which intersects the ridge from the north at 90 degrees, had been turned into a cascade of small fields, these flowing between copses where the ground was unploughably steep, or the trees too daunting in their size. A time-travel to Woodston in 300 BC discloses the farmstead as one of many such small farms along the Teme valley, each surrounded by its own fields, and in each holding both arable and pastoral farming practised, hedges disciplining and dividing the landscape.

The troubled weather of the Iron Age dragged troubled times behind it . . .

~

The Iron Age Scene at Woodston, 300 BC, The Frame and Canvas of the Picture: a decade or more back, I read the novel *The People of the Black Mountains* by Raymond Williams, a Communist Cambridge don. I loved it then; I love it more now, as the years have rolled, knowingness gathered.

Williams was born local to where we then lived, the Welsh border up above Abergavenny; he was a Pandy boy, Williams, son of a railway worker. The curious thing about Williams' novel, and the reason it left its indelible imprint on me, is its absolute negation of dry-as-dust Marxist historical materialism in favour of imagining the past from the position of geographical rootedness: to stand somewhere, and look back and see the images that have led to the here and now. (Curiously, the 'imaginative' historiographical method of the Marxist don is almost identical to that of the doyenne of conservative history dons, Dame C.V. Wedgwood, who was not unknown to me in my exile at various universities.)

The other day I stood on the linden ridge above Woodston, under the August moon, and looked down, and caught glimpses of the Iron Age scene, heard the babble of voices on the soft breeze. A girl . . .

> . . . leads two ploddy oxen, Gog and Gilgol, up the ridge to the farmstead, the wooden wheels of the cart squeaking like mice-chant. The cart, sides made high with ash poles, containing another load of wheat, brilliant white in the harvest moonlight . . .
>
> Her grandmother had smelled rain on the wind, so the family have been cutting the wheat from when the birds began singing . . .
>
> A dog Ruc, as big as a calf, is beside her, tongue lolling. She is glad of its company, although it is only a short distance from the three wheat fields at the bottom of the ridge to the farm, four hundred of her own strides. Raiders from the west had come last year, on horses, stealing cattle. They lived in mountain caves, and their grass had drowned in the rains. They were the 'Welch'. Stealers.

As the cart lumbers up the ridge behind her, the girl sees below her: all along the shining Trout River, scattered lights: the hearth fires of the stone houses with their doors open. Although it is August, and warm, the hearth fire never goes out. There is always the smell of woodsmoke in the valley.

The fires reflect the constellations of the stars. The sky is very clear, the land less so. Wood, more than three rivers' wide in breadth, separates the girl's home from the Dod people to the north, although she can hear, through the black trees, their voices as they too work cutting the wheat. The tree and the people fight for space.

From the wood between her and the Dod people, there is the grunting of swine; the slave boy keeps a watch on the swineherd, by night and day. Except when her father presses him to wield the mattock or spade to make channels, so the water runs off the fields.

The deepest of the night's shadows lie in the ditch encompassing the Woodstoni farmstead; it had taken the clan, all ten of them, a year of scraping with cow shoulderblades, axes and sticks to dig the ditch that Father said was necessary to make attackers stumble and fall. Her hands hurt then, even when bound thick

with wool woven from the puffs of sheep fleece
she had picked off the bramble.

On top of the ditch is the enclosing wall of ash
stakes, their tops sharpened with axes, gleaming
as wolf fangs. The girl enters the farmstead
through the palisade; beside the entrance is a pile
of stones gathered from the shore of the Trout
River below the house. They have brought up
the stones in buckets hanging from branches
across the shoulders; she had felt yoked then,
like Gog and Gilgol. The stones are for
slingshot. By the gateway there is also a tower of
spears, close packed like young growing trees.
Her father has an iron sword, which he sharpens
every evening by the hearth fire so it glints in the
style of winter ice. The sword is a thing of pride
to her father, as well as a thing of use. Only one
other family she knows of has an iron sword.
The Woodstoni are people of plenty.

Now inside the farmstead, the girl can see,
through the doorway of the stone roundhouse,
her grandmother keeping vigil over a pot on the
hearth fire, her body, framed by the doorway,
as perched and thin as a heron fishing. Her
grandmother's grey hair runs down her back in
rainy strands, and her face, caught by a sudden
plume of flame from a lind log, is the colour of
ripe nuts. The girl, at ten, already has skin

growing darker from labouring outside in the sun. A slave boy digs a channel with a mattock from the pen where the cattle are kept overnight, so their effluent runs away down the hill . . .

The girl leads the ox-cart into the open-sided barn, piled high with tree boughs, and begins unloading the wheat, the grains of which are food for humans, the stalks of which are fodder for beasts, although the longest of which will be woven into effigies to be sacrificed to the Hare Goddess so she ensures a good harvest next year . . .

~

Not, of course, difficult to imagine the above because, really, between the Iron Age and the twentieth century, what happened down on the English farm? My own mother, sufficiently 1950s soignée 'starlet' to win beauty pageants and be featured in the *Hereford Times*, led a waggoner's cart to Woodston, and was yoked herself, beam over her shoulders, to carry water, when the family lived away from Woodston on their own farm at Fownhope, one dry summer. In that moment she was all the women of the countryside before her.

The barn at my childhood house in the 1970s was nothing but stuck-in upright tree trunks with a plonked, cascading stone roof; and, I kid you not, the bin in which we kept the feed for the geese was raised up on boulders to prevent entry by rats – the exact system employed by the Iron Agers, who placed their granaries on stilts.

My grandmother, sitting beside the fire watching *Dad's Army* on TV, wove corn dollies (i.e. pagan sacrificial effigies, despite her being very orthodox C of E) from long wheat straws. This was 1977, which was the last good year. Afterwards, long straws were impossible to source, because all farmers locally – my grandparents being definitively retired – had switched to the dwarf wheat types compatible with chemical sprays. The towering old breeds of wheat could not, literally, stand the dousing with the herbicides, the pesticides, the fungicides, the molluscicides. Farmland birds went into freefall. It was the last year I saw corn buntings nesting in our village in south Herefordshire. On Radio Luxembourg, the Sex Pistols sang 'God Save the Queen', and prophesied that there was 'no future' in 'England's dreaming'.

You don't have to take my word for the decline in our farmland birds. You can take that of the Department of the Environment, Food and Rural Affairs, which published in 2014 'Wild Bird Populations in the UK, 1970 to 2014'. Regarding farmland birds the salient points were:

In 2014, the breeding farmland bird index in the UK was less than half (a decline of 54 per cent) of its 1970 level – the second-lowest level recorded.

Within the index over the long-term period, 21 per cent of species showed a weak increase, 21 per cent showed no change and 58 per cent showed either a weak or a strong decline.

Most of the decline for the farmland bird index occurred between the late seventies and the early nineties, largely due to the impact of rapid changes in farmland management during this period.

The long-term decline of farmland birds in the UK has been driven mainly by the decline of those species that are restricted to, or highly dependent on, farmland habitats (the 'specialists'). Between 1970 and 2014, populations of farmland specialists declined by 69 per cent while farmland generalist populations declined by 9 per cent. Changes in farming practices, such as the loss of mixed farming systems, the move from spring to autumn sowing of arable crops, and increased pesticide use, have had adverse impacts on farmland birds such as skylark and grey partridge, although other species such as wood pigeon have benefited.

Four farmland specialists (grey partridge, turtle dove, tree sparrow and corn bunting) have declined by 90 per cent or more relative to 1970 levels.

Difficult, I think, to see how the current mania for planting trees in Britain in re-creation of wolfy wildwood will improve the lot of the corn bunting, which requires open, arable fields.[6]

I should probably have confessed at the outset of this book that it is an unashamed defence of farming, and the English farmed landscape.

~

To deconstruct the above Iron Age Woodston scene with its half-dozen fields, of half an acre each, somewhat more.

The Iron Age roundhouse was sometimes partitioned, so that humans and animals aboded together. Aside from the prosaics of preserving the animals for their food and monetary value, the people found the animals' bodies, in the shared space of the house, provided warmth for body and mind. (Animals are companionable.) In all long-houses from primitive time onwards, the byre end is lower than the human quarters. (Effluent never flows uphill.)

Dwellings shared between beast and sapiens continued in neighbouring Herefordshire until the seventeenth century; we once lived in such a house in the Black Mountains. There were echoes inside those stone walls of that Black Mountain longhouse of the Old Time. Mice scuttered and skittled. The house mouse came with the prehistoric invaders, since it followed the frontier of civilization, so-called.

Even now, 'stockpersons' are always happy to sleep

with their animals; the groom with the horse, the shepherdess with the flock during lambing. I have bedded down in a Richardson twin-axle horse trailer with a sick pony to monitor an IV drip, lain down with the ewes on a hillside as they gave birth under March's starry skies, sat up in corrugated-iron sheds while pigs farrowed on Bethlehem straw . . .

People exterior to farming invariably believe the relationship between farmer and animals to be entirely master and servant, whereby the farmer is the master, the animal the servant, but often the farmer is the glad servant of his or her stock.

Today, I talked on the phone with my aunty Marg, born at Woodston in 1930. My grandfather wanted her and her four sisters, as he wanted me, to have a job 'in the dry', away from farming. Marg went for a year as a nurse, was ill with a flu-like illness the entire time. Then came back to Woodston, to look after the pigs, and was herself as happy as a pig in clover.

Then married a farmer, Tom Yarnold.

If ever in your life you touch farming, you never want anything else. Not really.

~

The Prehistorics were a meat-eating people; livestock numbers in the Bronze and Iron Ages were comparable with medieval times, and the Medievals loved flesh.

Humans have 'canine' teeth for the ripping of flesh.

The Prehistorics merely obeyed their biology, accepted a natural order where they were omnivorous, and understood that we are all beasts. So, I fret that today's vegans and vegetarians, who abhor the consumption of all flesh on principle, deny Nature and elevate themselves above animal appetites. And if we do all become vegans, what happens to all those breeds of farm animal that are our heritage? Hereford cattle, Soays, the Aylesbury duck . . . Churchill considered White Park cattle so intrinsic to British identity that he sent a herd to North America during the Second World War for safekeeping.

Plus: imagine a landscape without traditionally kept farm animals. Would it not be soulless? Boring. Dead.

People can be very cavalier about farmed animals. Thus, my own personal extinction rebellion is on behalf of the farm breeds, such as the Scottish Dunface sheep, that have disappeared off the face of Earth as decidedly as the Dodo. Did they not have worth, those farm animals? Is there a hierarchy of virtue, whereby Wild=Good, Farmed=Bad?

Ultimately, isn't #Meatfree ludicrously anti-ecological? Woodland that is *managed* – rather than running 'wild' – by allowing pigs in for pannage (nut gathering) and cattle for browse *increases* biodiversity. I once managed a wood in this way for four years, during which the tawny owl clutch size increased by 400 per cent and the wood warbler population by 200 per cent – just for example.

It's the insects, stupid. Where there is (organic) muck, there is the magic of a multitude of wee beasties. There are flies that adore cow shit so greatly that they lay their eggs in it as it spurts, warm and steaming, from the cow's arse.

~

To deconstruct the Iron Age scene at Woodston still more: when I was seven my grandmother took me for a walk along the lane at Withington and taught me to smell rain on wind; even then I was aware I was being inculcated in an art simultaneously ancient and demising.

The mattock was a farm implement as indispensable as the plough for nearly two thousand years; essentially a pick-axe with broader metal blades, the mattock smashed clods of earth to produce seed-friendly tilth, as well as excavating channels to allow drainage, of water, of effluent. (Been there, done that, so many times I have lost count, including this pluvial January, when the track flooded into the stables, meaning I had to channel out both water and effluent.) At Woodston, as everywhere in England, wet land was a barrier as definite as gravity. And Woodston was very wet.

There were spades before the Iron Age, their blades of flat wood, usually oak or ash; only iron, though, could be keened to slice into the earth easily and surgically. The same spade was used to dig the awkward corners the plough could not reach.

~

A hard lesson in reconstructing the Iron Age past, from last September:

Take a spade with an oak blade, 9 inches long on its bottom (business) end, keened.

Time taken to dig over a yard-square plot, two inches deep? Twenty minutes.

Take an iron-blade spade of the same dimensions.

Time taken to dig over a yard-square plot, two inches deep? Five minutes.

Robotics and algorithms are instilled and installed in modern farming; inside the tractor cab the computer screen indicates the cutting depth of the plough. But the plough is still fabricated from an iron derivative, steel.

We remain in the Iron Age, those of us who farm. We remain tied and indebted to the past. Thankfully.

~

Pigs were rarely brought inside the Iron Age farmstead. Aside from being the most troublesome of animals to confine, they were damn near impossible for rustlers to steal. As a veteran chaser of escaped pigs I confirm their resilience to being herded over distance. A rustler, of course, might have been tempted to slaughter the swine, and carry it away . . . but how? A pig weighs an eighth of a ton.

When our dear old Gloucester Old Spot pig Primrose was put down, she had to be dragged from the death-site, the top of a December-muddy paddock, down into the concrete farmyard to be collected by the knacker, Mr Harrison, and his trailer.

I tried using Willow, our miniature Shetland pony, to pull the dead pig, but equids do not like porcines, alive or dead.

The slope was too steep for a tractor to get up. The quad bike just skittered drunkenly sideways on the wet decline.

So I pulled the dead pig, blue nylon rope under her chest, downhill, leaving a slime trail in the mud and in the grass.

Even with gravity on my side, it took me an hour of slathering, sickening effort. At one point I had to tie the rope around my chest, a facsimile of the rope around the pig's chest.

QED. *Quod erat demonstrandum.*

~

The attitude of the Prehistoric people to the wildwood was ambivalent; on the one hand it was the place of cavernous dark in summer, steely and unforgiving in winter, the dangerous den of sharp-toothed beasts the year round. On the other hand, it was shelter from raking wind, the pigs' scavenging ground, the source of timber for buildings, fencing, fuel and charcoal. Oak and elm were used for the beams of ploughs and the runners of the sledge, the wheels of the cart; since metal was precious, it was used solely for the plough's cutting edges, and in time of war the metal was reshaped into swords and spears.

Perhaps above all the wildwood was the place to gather winter fodder for the stock. Hay – grass grown and dried to feed livestock – was only in the infancy of invention, for one supreme reason: grass needs a sharp blade to cut it, and sharp iron blades were scarce. As a rule of thumb only 4 per cent of lowland England in the late Iron Age was given over to meadow, the ground devoted to haymaking.

In hard winters in olden times, food for the farm beasts could be so lacking that, commonly, one in five animals did not survive. Tree hay, leafy boughs gathered in midsummer and dried, made a difference. To ensure plenty, the Prehistorics would have coppiced (cut off at ground level) or pollarded (cut off at shoulder height) suitable species of tree, meaning essentially they farmed the trees. As I say, the Prehistoric peoples would have seen the lines between farming and foraging, domestic and wild, as grey. Blurred.

Tree hay was also vital in drought; trees, with their deeper root systems and mycorrhizal fungal associations, can access moisture and nutrients and produce green leaves when the grasses have dried up. Tree leaves are known to have medicinal benefits and stock will self-medicate – what is technically known as zoopharmacognosy – where they have the opportunity. My grandfather Joe Amos told me in 1975, when I complained about my pony wanting to munch the elms in the hedge instead of gallop, 'Let him be, he's getting his vittles.'

From the horse's mouth, translated by Poppop, I learned animal intelligence.

~

An experiment in making tree hay, France, a summer ago.

Under a spreading lime tree. Outside the sanctuary circle of shade, only heat haze and a dissolving world.

Despite the belly-dancing of the air beyond the tree, there is a stillness to the mid-afternoon, as if I were stuck in a glass jar; the sole sounds are the hacksawing of cicadas in the walnut orchard and the equally rhythmic, metallic clicking of tree-loppers.

I am cutting the lower fronds of the lime tree – the large-leaved variety, *Tilia platyphyllos* – my ladder propped against the trunk; this, at 2.8 metres in girth, has the stature of a Parthenon pillar. But it is deeply creased, in the manner of old elephant's skin. Up and down the bark's cracks scuttle firebugs, brilliant and scarlet. The tree seems to seep blood.

The lime is a pollard. The three main branches atop the trunk have been cut repeatedly over the decades to produce Medusa masses of whippy branches and a canopy so dense it permits no poetic shive-light or romantic dappling. Under the lime tree, the shadow is seamless.

No ornament, my lime. (Although it is hermaphroditically attractive, with its masculine frame, its feminine frilly bathing-cap foliage.) Like the other limes planted in remote Charente in the nineteenth century, it was intended as a multi-tasking working tree: to supply the house with shade, that welcome gloom, in summer, and to furnish wood for the home fires in winter. The same wood, easy to work and hard to split, would have been carved for toys and cut for poles.

On and on go the historic French uses of *le tilleul*. The flowers fed the bees, flavoured drinks. In Alain-Fournier's 1913 novel *Le Grand Meaulnes*, a requiem-in-advance for lost idealism (the author himself was killed in the Great War), the schoolboy narrator recalls attic rooms 'where we kept drying lime leaves and ripening apples'. These leaves were for soothing lime tea for the mind, and a more soothing still medicine for gastric ailments.

The last significant use of lime's heart-shaped leaves was animal fodder, which is the reason I am chopping boughs of the tree on this July day. In ancient northern Europe, cut and dried arboreal leaves were stored for feeding to the animals during winter; in south-west France the hunger time is droughty, oncoming August.

Pruning for tree fodder is the worst of jobs in farming, the best of jobs in farming. Looking up into the tree's depths for the right place to cut is a literal pain in the neck. Neither does my protective clothing work. In a nod to Health and Safety, I am wearing the farmer's traditional head guard – a tweed cap – and in the need to repel insects and heat, a white shirt, buttoned to the neck and the wrists. So, of course, all falling branches jink to avoid my head and stab my shoulder, while all the biting insects settle for the exposed rim of my neck, leaving me with a pearl-choker of bites.

Entomologists, when desiring to discover the insect life of trees, beat the latter above a collecting tray. A white 100 per cent cotton shirt is every bit as competent in the display of insect life. Aphids, ants, firebugs, wood lice, beetles, nameless black flies, nameless green flies all shower down on me, though nothing quite makes my skin crawl so much as the caterpillars of the lime-hawk moth. Normally nocturnal, the caterpillars are motioned into afternoon activity by my pruning. They grow to 6 cm in length. They are grotesquely tumescent, ever ready to burst. They are glaring green – with a curved neon blue horn at the tail end.

Under my lime, it is cell-like and tranquil, a place fit for a monk or a scribe or a philosopher. (Ask Isaac Newton why being beneath a tree coaxes thinking.) My farmer's thought for today is that the contemporary arborists' wisdom that the society of trees is a hippie, communistic utopia (the 'Wood Wide Web'), in which the trees have personhood like the Ents in J.R.R. Tolkien's *The Lord of the Rings*, is plain false. It is to impose on trees, by the feeble pinky-grey cells of the human brain, characteristics they do not really have; a sort of species imperialism that diminishes them, and us. Trees are trees, and none the less for being so.

This is the last batch of dried-leaf fodder for this summer. I have waited for the wood pigeon squab in the lime to fly its nest-raft, the black redstart young to quit their hole along a cranny, behind a gnarl. I have made tree hay from other limes, from maple, from ash; there is about a ton of the stuff hanging up in 'bales' in a shed to dry.

Today, in this fierce heat, the leaves are drying within hours of me simply throwing them out into the sunlight, which sears in their minerals and nutrients. Their goodness.

Later, when carrying the lime boughs by the armful through the white light to the shed for storage, I launch one towards my horse, Zeb. As the branch arcs the pure blue horizon it leaves a trail of scent; it is the same fabulous smell as a packet of China tea, fresh opened.

Zeb gobbles the sun-baked lime leaves in preference to meadow sward. Thus, the ultimate proof of the utility and tastiness of tree hay comes direct from the horse's mouth.

~

The wildwood, of course, is a Modern Romance: it was never truly wild; the beasts husbanded it unknowingly, with their mouths and their feet; they let in light, and churned the surface to allow woodland plants to germinate and colour the scene. In the glades created by the aurochs, plants of the tundra of late-glacial Britain, such as plantain, found purchase and sanctuary, and these would become mainstays of the pastureland of the Prehistorics.

Whatever remaining uses the Prehistorics had for the

wildwood, the impetus of English farming was to put ever-increasing amounts of land under the plough. More land and a higher cereal yield allowed the population of England to increase in excess of a million.

Cultivated wheat. Love it, hate it, it is the undisputed people-grower. The dominant wheat cereals of the late Iron Age and Romano-British periods were emmer (*Triticum dicoccum*) and spelt (*Tr. spelta*).

Iron Age farmers knew their land, their job, their crops. Their cereal yield was, at the minimum, a ratio of seven grains harvested for every seed sown, and the leftover straw, threaded with 'weed', made highly palatable, nutritious fodder for the livestock. Fertilizer inputs were low for the simple reason that emmer is not nitrogen-thirsty like modern cereal hybrids. Spring-sown emmer and spelt, when established and aided by hoeing and hand-weeding, compete successfully with arable weed infestation, including the English farmer's nightmare, charlock (*Sinapis arvensis*), the oh-so-pretty yellow serial strangler of cereal crops across the ages. Also, the thick, large 'spikes' (seed heads) of emmer and spelt are nearly impervious to bird attack.

In other words, Iron Age farmers were getting off the land as much nutrition, when input is compared to output, as modern intensive farming. Without degrading the soil.[7]

Agricultural progress? So frequently a chimera, a politician's promise, a chemist's stunt. We have lost fields of glory.

Farming changed the flora of the isles, away from predominantly woodland species to those of the field, whether pasture or arable. When the Iron Agers took their scythes to the wheat on acid soil such as Woodston's, they travailed amid a riot of floral hues. There were thistles (*Cirsium spp*), fat hen (*Chenopodium album*), poppies (*Papaveraceae*), charlock (*Sinapis arvensis*) and hedge mustard (*Sisymbrium officinale*), red shank (*Polygonum persicaria*), sowthistle (*Sonchus arvensis*), and pale persicaria (*Polygonum lapathifulium*).

What birds sang then, when the Iron Agers bent to their fieldwork? The Prehistorics changed the soundtrack of Britain from predominantly woodland to mainly farmland. Literally, the Prehistorics created the open space for our national bird in music, the soundtrack of English spring until the 1970s: the ascending skylark.[8]

Skylarks do not silver-coin trill over woods. Neither do curlews ghoul-wail over forest. Neither do lapwings 'peewit' for spring over copses. It was the cutting down of the wildwood for agriculture that gave the skylark, the curlew, the lapwing, their advantage.

~

I stopped off yesterday evening at Lindridge, on the way to dinner at the Talbot Arms, for the view over the Teme at sunset. The soil was warm with spring. A peacock butterfly floated by on the balmy 7 p.m. air, landed and nectared on a cuckoo flower. Pansy-pigmented, the butterfly added its colour to the scene. Always, the more the flowers, the more the butterflies.

A companionable meadow pipit flitted here, there.

The sward was tunnelled through by voles, and kestrels sat around on the nearest trees, flame-tapers in the sunset. By and large, peacock butterflies, yellow rattle, meadow pipits, field voles and kestrels do not inhabit woodland. Farming gave England a particular pasture-land ecology, as well as an arable one. But 'farmland' both.

~

Once more, for the avoidance of doubt: the Iron Age was already the Modern Age, the lines of continuity between then and now direct and immaculate. The plough behind my Ferguson tractor is no different in principle or construction from the Iron Agers' ard towed by oxen. My chain harrow no different from the spikes set in wooden frames they used for harrowing; their pens of wattle-and-daub or woven hazel no different from my galvanized sheep 'hurdles'. Grain pits and kilns are nothing more than prototype modern siloes and grain dryers.

Such was the progressive, extraordinary England the Celts invaded in 300 BC: a patterned landscape of fields, arable and pasture, with islands of woodland and efficient farming practice. The Celtic times were good times for English farming, and for 150 years before the Roman invasion the land produced such a bounty there was product to trade abroad. Strabo, the Pontine geographer born in 64 or 63 BC, classified the exports of Celtic Britain as 'corn, cattle, gold, silver, iron, hides, slaves and clever hunting dogs'.

The Celts worshipped cattle, as well as bred them, a reminder that the mind of the prehistoric farmer differs from that of the industrialized farmer of the twenty-first century, being less utilitarian and instrumentalist. For the Celts, animals were beings, not products; and the land, the stones, the wind, the sun, the rain implicit with gods and spirits.

White cattle with red ears, the Celts considered, rather charmingly, came from the Otherworld, where they belonged to the Sidhe, fairies from another dimension who kept herds of them under water. Occasionally these fairy cattle would appear on the shore as a gift for a human farmer, and if treated with respect would provide them with boundless milk.

Celtic white cattle live on in the feral herds at Chillingham castle in Northumberland and the Duke of Hamilton's Cadzow Park in East Lothian. Domesticated versions are the horned White Park and Vaynol, and the polled British White.

They are flesh and living prehistory. Not all the clues to the past are buried in the ground.

~

The Celts' white cattle were legendarily the animals of Druidic sacrifice. Pliny writes about a Druidic sacrifice of two white bulls in AD 77:

The Druids – for that is the name they give to their magicians – hold nothing more sacred than the mistletoe and the tree that bears it, supposing always that tree to be the oak ... The mistletoe, however, is but rarely found upon the oak; and when found, is gathered with rites replete with religious awe. This is done more particularly on the fifth day of the moon, the day which is the beginning of their months and years, as also of their ages, which, with them, are but thirty years. This day they select because the moon, though not yet in the middle of her course, has already considerable power and influence; and they call her by a name which signifies, in their language, the all-healing. Having made all due preparation for the sacrifice and a banquet beneath the trees, they bring thither two white bulls, the horns of which are bound then for the first time. Clad in a white robe the priest ascends the tree, and cuts the mistletoe with a golden sickle, which is received by others in a white cloak. They then immolate the victims, offering up their prayers that God will render this gift of his propitious to those to whom he has so granted it. It is the belief with them that the mistletoe, taken in drink, will impart fecundity to all animals that are barren, and that it is an antidote for all poisons.

It is not hard to envisage the oak groves at Woodston, on top of the linden ridge, with hanging mistletoe, in the way that apple trees hereabouts are nowadays baubled with the plant. This area is the mistletoe capital of England. Tenbury Wells has a mistletoe sale every Christmastide in the old Round Market. Buyers today come from as far away as Ireland. The trade lines and trade lanes of the Woodstoni continue into today.

CHAPTER III

HEED THE PLOUGH

*From the Roman occupation AD 43 to the Norman
Conquest 1066 – Above all, the Anglo-Saxon centuries –
Agricultural time is not linear time, or political time; it has
other measures and scales – It was the coming of the heavy
plough, pulled by eight oxen, to Woodston circa AD 850
that enabled the heavy, fertile soil beside the Teme,
Woodston's best red soil, to be broken*

VENI, VIDI, VICI. The Romans came, saw, liked what they
saw, and eventually conquered the woad-daubed Celts.
(The faces of the battling Celts would have looked like
those of Six Nations rugby fans, also painted in tribal colours; history is ever enslaved to repetition.)

Famously, the Romans, from their occupation in AD
43 to the withdrawal of the Eagles in AD 410, turned
lowland England into an imperial breadbasket, putting
as many as four million acres under cultivation, plus cattle and sheep pastures. They exploited lowland England's
farming potential to the hilt, harvesting as much as 15–20

71

bushels of corn per acre, or over 1,000 lbs in less *recherché* speak. Wheat was the dominant cereal; the main species were spelt (*Triticum spelta*), emmer (*T. dicoccum*) and the compact bread wheat *T. aestivo-compactum*. Barley was grown frequently, though less frequently than wheat. Rye and oats were scarcer, save in the wet west and north. Under the Romans the population of Britain rose to 1.5 million.

What did the Romans ever do for farming? The Romans knew their manure, and how to improve soils by applying farmyard faeces. Sings Virgil:

> *Land by flax and oats*
> *Exhausted, is consumed where slumber-steeped*
> *Poppies of Lethe lift their glowing heads.*
> *Naytheless, by change of crops is labour eased;*
> *Only be not ashamed with fattening dung*
> *To enrich dry soil, or, when life is gone,*
> *To foul your hands by scattering wood-ash.*
> *So will rotation rest your land: meanwhile*
> *There is no loss through acres left untilled . . .*

The Romans recognized the benefit of animal dung to the land, if ignorant of the chemical processes involved. Animals penned on land improved it, as did excrement spread from overnight or over-wintering enclosures; truths half glimpsed by the Prehistorics, but crystal-clear to the Latin invaders.[1]

The Roman era is surprisingly modern. The Roman rich lived in grand villas – making them the originals of that enduring species, the English country gentleman – and read improving volumes by agronomists Cato, Columella and Pliny the Younger on costings and maximization, making them also the original farm managers, the occupation of my grandfather. The estates of the Roman magnates produced wheat and wool for local and foreign markets, sold via the offices of merchants, through the mechanism of money: proto-capitalism.

The Romans wafted the velvet glove as well as wielding the keen-edged sword. Under *Pax Romana*, beyond the walls of the estates the people were left to work untroubled in the ancestral fashion, the Romans satisfying themselves with taking a cut of the proceeds, because for the godless Romans – the most matter-of-fact of peoples – money was deity. The Roman tax inspector ensured a tenth or one-twelfth rate for taxation, although 'resistant' areas were assessed at a penal rate nearer three-fifths. In docile parts of the country, like Lindridge, the quota was probably collected as rent or share-cropping.

The closest Roman estate to Woodston was sixteen miles away at Droitwich. Otherwise, the celebrated Roman villa was conspicuous by its absence. Roman military power reached Lindridge in about AD 52, and it must be assumed that the native aristocrats and peasants either professed allegiance to Rome or were replaced. Compared to the rest of England, the conquest had relatively little effect locally. The Roman villa captures the imagination, but only 620 are known in Britain, and few could be described as 'des res' or 'de luxe'; most were farmhouses laid out on the 'basilical' plan, with a long hall divided into nave and aisles by two parallel rows of posts, or even a crude two-room dwelling, one part of which housed the family, the other the livestock. In effect, this was the same shared accommodation known to the Iron Age Celts.

Woodston was on the western edge of what anyway was a peripheral province of the Roman empire, one habitually referred to as 'a country of setting sun, remote from our world'. Lindridge and the isolated, enclosed 'native' Celtic farmsteads continued much as before. The Romans barely scratched the surface of life at Woodston, where the peasant farmer grumbled, paid his taxes, had no gratitude for *Pax Romana*. A divine camera, watching Woodston on high, would have recorded the same annual film loop for almost four hundred years.

Almost. This is not to say that the Romans were absent figures in the viewable or journeyable landscape; there are remains of a Roman brick kiln at Mamble (2 miles), traces of Roman habitation at Sodington (ditto), and Roman forts at Clifton-on-Teme (6 miles) and Wall Town (7 miles). It may also be safely assumed that the Woodston farmers took advantage of Roman agricultural innovations and inventions, aside from more systematic manuring. These practices had a lasting impact on Woodston, if not immediately, at least in the long term.

The Romans were big on sheep; commercially minded, they responded to the demand for British woollen cloth, sought after in the emporia of Rome itself. The Neolithics introduced sheep, the Bronze Age people made them numerous, but it was the Romans who granted them their ascendancy. The Romans, who had little taste for mutton,[2] unlike Middle Eastern elements in their Legions, imported a hornless, white-faced short-wool sheep that, it is generally accepted, is the ancestor of the medieval short-wools, such as the Ryeland. The Ryeland was responsible for the fleece on which the entire Tudor cloth industry was based.

The Ryeland was much bred and propagandized by my grandmother's ancestors, the Parrys, who were flunkeys at Queen Elizabeth I's court. They legendarily gifted Gloriana soft stockings of Ryeland wool, after which she declared she would wear no other on her virginal legs.

~

The Romans, on their own table, preferred a ration of beef. The roast beef of Merrie England is an old, old dish, the taste for which has been acquired by every invader. The preponderant race of cattle in Romano-British time was the old *Bos longifrons/brachyceros*, although remains of larger animals indicate cross-breeding with imported Roman cattle. Genetic manipulation is an ancient, black art.

What else did the Romans ever do for Woodston? The Roman Army arrived with well-equipped cavalry units, and horses that were the product of many generations of controlled breeding. The blood of all the races of horses in the ancient world then went into the British studs.

Our earliest ancestors ate horses, which were indigenous to Britain, a quarry of the hunter. Neolithic people, in whose settlements the remains of horses are frequently found, ate them, probably milked them, but did not ride them. By the Bronze Age, horse 'tack', bits and bridles, are found in tombs of the great, indicating that the horse was ridden by the elite and, perhaps, the warrior, the proto-cavalryman. (The art of horse-riding began on the steppes of central Asia or south-east Europe and was a cultural import to England; there was no island mentality. There was continuous communication with 'the Continent'.) Later, the British harnessed equids to chariots, as did the ill-fated Boudicca of the Iceni, a notorious horse-trading (read: horse-stealing) tribe, who, according to legend, donated us the noble beast that would be known as the Suffolk. It did not occur to the British to yoke equines, despite the abundance of evidence before them. (The Romans had heavy horses.) At Woodston, the first employment of the horse – a shaggy *Equus agilis* or *Equus robustus*, both of which resembled a New Forest pony – would have been to round up free-ranging cattle herds bred for meat. My grandfather used a horse for checking on the stock, as do I.

The Romans also introduced to these isles, inter alia, the vine, the fig, the plum, the mulberry. They imported the cherry and the cultivated apple. When Caesar invaded, he found the Celtic inhabitants fermenting the juice of the native crab apple, *Malus sylvestris*. The Romans called the drink *sicera*. Cider. The bitter taste was not suited to the Romans' sophisticated palate, thus they introduced the sweet apple, which itself became adopted as a cider apple.

Without the Romans, Woodston would not have enjoyed its banks of blossomy cherry and apple orchards – seven of them by the 1850 field map. Carters Orchard, Pool Orchard, Barn Orchard, Fold Orchard, Upper Bowcroft Orchard, Bowcroft Orchard, Jones Orchard. Tenbury would never have been titled 'The Town in the Orchards'; there would have been no pink-blossomed spring days, no ripe-scented September mornings of my mother's memory.

The cider latterly made on Woodston Farm, as on all Herefordshire and Worcestershire farms, wasn't just for the farmer and his kin: it was part of the workers' wages, amounting to 2 quarts (2.25 litres) a day for a man and 1 quart (1.15 litres) for a boy in the nineteenth century. Poke about an old hedge hereabouts with a stick and you are sure to find a dispensed-with earthen jar, from which the farm labourer swigged during a rest from haying in the medieval, Elizabethan, Hanoverian, Victorian sun.

In the 1950s my grandmother's cider was sufficiently

good for her to decide she was a born publican and, for a while, she became the landlady of the Black Lion, as well as the farmer's wife at Woodston.

The sweet chestnut also came with the Legions who, in the style of southern Europe, made its fruits into flour. The English decided it was a savoury, the stuff of stuffing, and eventually the base for the Edwardian country classic chestnut soup, and something to be roasted over an open fire at Christmas. Or in the Rayburn stove (the proper country person's AGA), which is what my Amos grandparents did. The sweet chestnut did not take everywhere, but it did take in west Worcestershire, at Woodston.

Another Roman innovation was the cat, as was the domestic goose (my grandfather's favourite Christmas roast); the pheasant and the black rat, originally a native of India; the chicken, much boosted by the Legions, although first brought to these shores by the early Iron Agers, who considered the fowl a god akin to Mercury (and buried them whole, as they did hares, which were associated with the goddess of fertility). In *De Bello Gallico*, Julius Caesar infamously claims that: 'The Britons consider it contrary to divine law to eat the hare, the chicken, or the goose. They raise these, however, for their own amusement and pleasure.' This is a pot calling a kettle black: the Romans were fanatically devoted to cockfighting. Archaeological evidence from Roman Britain shows chicken specimens possessing developed spurs bred for battle. The Romans liked a scrap: their military triumphs were predicated on a willingness to engage in bloody, close-up combat. They considered fowl to be egg-layers as a second resort. Blood before eggs.[3]

~

Anyway, if I cannot say that the Romans are my personal taste, I do appreciate living in their orchards, with their improved chickens, so that one enjoys such scenes as this:

At home in Herefordshire, if there was no hill in the way, I could see Woodston. It is one of those September mornings when the sky is scraped clean of cloud. The air is vital. The dew on the orchard grass is lubricious, like birth-fluid.

'Orchard' is perhaps a grandiose term for sixteen half-grown, half-standard apple trees on a quarter of an acre. The chickens, though, are content with the place. Hens, tamed junglefowl, remain jungly. Under a tree is a chicken's safe space. Aerial predators require a clear run; trees prevent this. The reason chickens loiter around the sheds in many free-range systems is the absence of arboreal cover in the 'range'.

Fear of the hawk is indelibly imprinted in fowl; it runs through them like lettering in Blackpool rock. I am thinking exactly this, as I fork litter from the hen-house floor through the door into the wheelbarrow, when the local sparrowhawk comes in over the treetops, a grey blade twisting into my bucolic scene.

The siren wail of avian alert goes up from the hedges; blackbirds mimic those 2 a.m. car alarms no one can turn off.

The chickens crouch, flatten under the cover of the apple trees. The hawk reappears, higher now: the approach dive, the killing time.

The hawk slashes down through the air . . .

Then she pulls up; there is no flight path through the tangle of twigs, the barricade of branches, the lattice of leaves. She flashes overhead; her pupils are obsidian, her talons curved graphene. She is murder on wings.

I like birds of prey, just not in my orchard, seeking and destroying my hens. So I hope the sparrowhawk saw the light of farmer's triumph in my eyes. The Bible basic of husbandry is that sheep may safely graze, hens may safely peck.

Predator panic over, I return to the foul job of cleaning up after fowl. Shifting shit, along with acting as minder, is the other indispensable of agriculture. It's all glamour down on a farm.

Out among the fruit trees, the chickens resume their relentless pecking at potential iotas of food as though the sparrowhawk had never been. The grass is high and wet, the autumn flush, and the chickens sail along sweetly on their hulls. Our chickens are a careful miscellany of different providers of shell colour: Minorcas (white eggs), Araucanas (greeny-blue), and a couple of Warrens (brown) chosen for their eggy plenitude, or high 'Hen

Housed Average' in the jargon. Of course, no flock is complete without Light Sussex, those white-feathered galleons with the big beam. Our Light Sussex lay a fetching pink egg, almost the tone of sugared almonds.

Some apples have already fallen from the trees. Scarlet and bright, the windfalls swarm with gaudy wasps in cameos of Bacchanalia. In the eye of the chicken there is no compassion and this is the vice that makes them effective pest-controllers. Apples ridden with harmful grubs usually drop first; the chickens cleanse the bad fruit, reducing insect pressure for next year. Orchards love chickens.

There is a manic clucking from near the gate, five chickens fighting over a writhing morsel, until one hen breaks into a thigh-pumping run and gets away with the prize.

Nature is red in beak and claw. Even domesticated Nature. Once our chickens mob-attacked a grass snake and hammered it to death with their bills. A sort of hysteria possessed them that day; the snake was a yard long, and struck back, but the dense down on the chicken's breasts rebuffed the snake's fangs. The hens pecked on and on until the snake had no more life than a cut green stick.

Chickens, as I say, are from the jungle, via Rome. Civilization is just skin deep. In fowl, as in humans.

~

They had their novelties, the Romans: their whetstone-sharpened, two-handed scythe, which finally made an efficient close-cut hay harvest; the heavy plough with a coulter; wells; the rotary quern (previously flour was produced laboriously between a stationary lower stone and an upper stone rubbed backwards over it); the water mill; the iron spade, which would have had an important impact on field drainage. Their harrow, a heavy sledge studded on the underside with stumpy projections of iron or wood or flint, disturbed the ground sufficiently to be called a *tribulum* – from which is derived our word 'tribulation'. In their kilns, the Romans installed hypocausts, a centrally heated drying system Woodston would not match even in my grandfather's heyday.

The Roman occupation provided Britain with what it had hitherto lacked, an urban population requiring farm produce. For the Celtic-Britannic farmers of Woodston, all roads led to Wall Town Fort, just up by Cleobury Mortimer, and the garrison there. Every farmer likes a customer, and customers now handed over coins, rather than bartering goods.

If the taproot of agribusiness goes deep back in time, so too does state interference in rural affairs. Barbarian invasion, over-taxation and inflation brought the glory of Rome down, as did state pressure to produce winter wheat, which reduced the stubble available to graze the stock, which in turn led to decline in meat production. During the latter part of the occupation, the weather – the curse and the beneficence of English farming – changed. Wet England got wetter, meaning heavy loams became more difficult to work. Consequently, the Romano-British rural system broke up after the fleeing of the Eagles in AD 410.

The Romans' greatest effect on the countryside was nihilistic; it was to begin its destruction, by founding the town. Although the Roman cities were not large, ranging from 100 to 200 acres, they housed, as well as citizens, tradespeople and professionals. The English disconnect from the land had begun, along with the occupational specialization that would put the power of agriculture into the hands of others, from mill-owner to veterinary surgeon. Urban sprawl commenced with the building of Londinium.

Even so, when the Eagles departed England, tracts of natural forest remained. The Kent and Sussex Weald stretched for 120 miles. The Forest of Dean still had the feel of wilderness.

It was the agriculturalist Anglo-Saxons whose mass invasions from the fifth century AD onwards completed the transformation of the English landscape; they fixed boundaries of field, woods and parishes still extant today. The paths the Saxons wore into the ground between home and field are our public footpaths. In ninety-nine cases out of a hundred, the English village today stands where it stood when the Saxons lost the Battle of Hastings in 1066.

By the time of the Norman Conquest, the open pattern

of our modern countryside was established from shore to shore. This included Woodston, finally and wholly wrested from the wildwood.

~

They came in their oaken longboats, from Germany over the North Sea, which they called jocularly the 'gannet's bath', but this watery traverse done, they showed no adventuring spirit whatsoever.

The Anglo-Saxons: they could drink (ale, fermented juice of barley being their alcohol of choice), eat beyond repletion when the chance allowed, and they could fight. But above all they liked to farm the land, the *terra firma*. I know because my fair-haired, blue-eyed maternal grandfather had a slave name, Amos, this imposed on the Anglo-Saxons by the Normans in 1066, and no one could stop Joe Amos farming. He was at it in his seventies.

These are the years when England and the English, a deeply rural people, were made. Our residual distrust of the forest, our cultural preference for long vistas over rolling acres, is inherited from the Anglo-Saxon invaders. They gave us our landscape of the mind, as well as our country's very name, *Engla-land*.

The Angles had no appetite for manufacturing, neither were they drawn to the Roman towns. The Saxon eighth-century poem 'The Ruin',[4] a portrait of Bath, begins:

Well-wrought this wall; Wierds broke it.
The stronghold burst . . .
Snapped rooftrees, towers fallen,
the work of the Giants, the stonesmiths,
mouldereth.
      Rime scoureth gatetowers
      rime on mortar.

Literary critics muse on the wondrous poem's meanings. Lament for the effect of time on man-made things? An acknowledgement that all beauty comes, ultimately, to dust? The most obvious point of 'The Ruin', however, is that the Anglo-Saxons resiled from doing up the place and reinhabiting it.

The Anglo-Saxon invasion was the largest population change in English history; there was some ethnic cleansing, with the Romano-British slaughtered or driven west, to end up in Wales, or Cornwall, which was much the barren same. In some places, in the 'Dark Ages', the Christianized Romano-British and the pagan Anglo-Saxons lived side by side.

I suspect, however, that the Romano-British farmers of Woodston were not among those moved on at the point of a sword, to go west into the mountains with the other remnants of ancient races; once, when I walked up the lane at night, I felt for certain that the small people had died there, and their blood was among its fertilizers. You would not give up Woodston easily, if at all.[5] (An intuition, no more: Pensax, bordering Lindridge, is composed of the Celtic *pen* for hilltop and *sais* for Englishman, suggesting the compromise of miscegenation, even perhaps a welcome mingling. Nearby Menith Wood also bears a stubbornly Celtic-Welsh name, *mynydd* being a mountain.) The earth of England is the graveyard of races, the one dead race upon the other, in layers.

~

The Anglo-Saxons were folk who followed a warrior chief who, on arrival here in *Englaland*, laid down his sword, divided his land among his followers, and began to pasture his livestock and to plough amid the existing forest clearings. Most often the resources of his warrior farmers were too small to enable them to tackle the work alone, so some cultivation and stock-pasturing were practised in common.

The Ango-Saxon farmer who took over the Romano-British farm of Woodston paid allegiance, in service or in kind, to his chief, in return for protection against murderous rustlers and Viking raiders. The obligations on both sides descended through the centuries to the heirs on both sides. The Normans did not invent feudalism. They bastardized it. The Anglo-Saxon thane or *thegn* was the prototype of the Norman lord of the manor who, in turn, was the precursor of the English country squire.

Despite owing obligations to his *thegn*, the Saxon

farmer was a free man, and thought of himself so. He was a proper peasant.

~

It is in the Anglo-Saxon years that Woodston enters history, receives its name: 'Woodston' is Anglo-Saxon for 'farmstead at the wood'. And the field names of the farm tell their story. At Woodston one field name is 'Ox Leasow', *leasow* being near pure preserved Saxon for grass keep; the ox, of course, was the Anglo-Saxons' primary motive power after they turned Christian and were therefore forbidden to keep slaves.

Like the Neolithics, the Anglo-Saxons spread via the river systems, but unlike those first farmers the incomers did not head upward, to the linden ridge, to the more easily worked but less productive, now much used, soil of the hilltop. They had no need, because they had the transformative implement, the heavy plough.

For two or three centuries after the arrival of the Saxons at Woodston, maybe nothing changed. The pace of rural life is unbelievably slow to city moderns.

The Anglo-Saxons took on the woods, but their felling of the trees was piecemeal, premeditated, systematic; a glade created here by the removal of a choice oak, a trackway hacked there for access. They were not, as arborists of the twenty-first century have it, tree-murderers, engaged in a final botanical solution, where it is always 'wildwood destroyed', not 'farmland gained'. If anything, the Anglo-Saxons resisted felling the trees on the linden ridge because the fertile lowland was their priority. So they moved south towards the river, and east towards the millstream.

But by AD 1000, at the latest, the Saxon heavy plough allowed Woodston's richest but wettest soil, dense as beef, running down to the river, to be opened up. The Anglo-Saxon heavy plough, as influential in its era as the Benz internal combustion engine or the NASA space rocket, had a mouldboard which turned the earth over, burying weeds, breaking up clods and improving drainage, plus a coulter in front of the ploughshare, and a wheel, making it little different to the plough on the arable field today. It was drawn by teams of oxen, two, four, even eight strong. (The little ard of the Celt needed just two beasts.) Late Saxon illuminated manuscripts are full of pictures of the heavy plough, because the monkish writers recognized the technological source of their own wealth.

~

The development of the plough is the most vexed question in agricultural archaeology; it is within the bounds of feasibility that the last tribe of the Celts to arrive in England, the Belgae in the first century AD, brought the heavy plough equipped with a mouldboard, coulter and wheel. What is certain is that the Anglo-Saxons possessed such technology, and it is they who radically changed the geometry of fields. With the heavy Saxon plough, there was no need for cross-ploughing, and every inducement for the lumbering team to plod on as far as possible before turning. So fields became rectangular.

The Saxon heavy plough is a Marxian dream, a seeming proof that technology is the driver of historical process. The time of Bede was a time of revolution; cereal cultivation boomed (especially free-threshing wheat, to the detriment of the old spelt), and winter planting of barley became possible; the oxen-hauled Saxon plough could even turn wet November ground. The plough also facilitated the famous Anglo-Saxon strip-farming system, whereby large open arable fields were divided into lengths, each about half an acre (a day's ploughing). These were divvied out among the local farmers, who all planted the same crop at the same time on (generally) a three-course rotation, the third year the land going fallow for recuperation.

The system came late to Woodston and Lindridge, as late as the ninth century, and was never anyway fully adopted. Progress at Woodston was always piecemeal, ad hoc, until the Victorian entrepreneurs arrived. Nevertheless, the heavy plough changed the skin of Woodston, changed the skin of England. The plough influence of the Anglo-Saxons is still evident at Woodston; the fields Shop Bank and Stephens Orchard retain their rectangular regularity to this day.

The heavy plough made an indelible mark on language too. We still *speke* Anglo-Saxon when working the land: a 'furlong', one-eighth of a mile, is a 'furrow-long'. (The furrow, when downhill, as in Shop Bank, allowed drainage, acting as a miniature dyke.) An 'acre' is the amount of land a man with two oxen could plough in a day, a square with sides just under 70 yards long.

The word 'plough' appears to derive from the Saxon *plou*, although the ultimate origin is unknown. Max Müller in *Science of Language* connects the word with the Indo-European 'to float'. The same word would be applied to a ship 'ploughing' through waves.

The ploughman became the aristocrat of the field, because his work was skilled, arithmetical, geometrical. For more than a millennium the ploughman would rule the furrow waves, with his own guild and religious rituals. The Old English blessed fields; conversely, community transgressors undertook the fire ordeal of walking over heated plough-irons. From Saxon times rituals directly associated with ploughing were incorporated into Christianity, and these looked forward to the Middle Ages, by which time 'Plough Monday', a celebration suffused with pagan fertility symbolism, emerged as one of the major seasonal festivals of the Christian calendar. The first Monday after the Twelve Days of Christmas, Plough Monday, was when agricultural communities restarted toil after their festivities.

Then there were the songs of the ploughman, passed down the generations, by speech and by quilled word. The very anthem of plough songs, the most famous of them all, is 'The Painful Plough', printed at Darlington in 1774. The folklorist Cecil Sharp noted that 'The adjective "painful" is, of course, used in its original sense of taking pains, careful, industrious':

> *Come, all you jolly ploughmen, of courage stout and*
>   *bold,*
> *That labour all the winter through stormy winds*
>   *and cold,*
> *To clothe your fields with plenty, your barnyards to*
>   *renew,*
> *To crown them with contentment, that hold the*
>   *painful plough.*

The ploughman of the song continues: 'no calling I despise/For each man for a living upon his trade relies', adding that gardener, merchant, all are dependent on the painful plough which prepares the land for the cultivation of food:

> *I hope there's none offended at me singing this,*
> *For it was ne'er intended to be ta'en amiss;*
> *If you'd consider rightly you'd find I speak it true,*
> *All trades that I have mention'd depend upon the plough.*

The words are from 1774; the sentiment belongs to 774.

∼

The heavy plough and the ox were a virtuous romance, like horse and carriage, love and marriage; because the heavy soil cleared by oxen-drawn plough was the sort of soil that could support them nutritionally, either in winter fodder, or in aftermath grazing when the hay was cut. There was a large increase in the percentage of cattle in Saxon times, and the laws of the Saxon King Ine of Wessex (688–95) show cattle farming to have reached a stage not unlike today, with the beasts contained within fields rather than free-ranging in woods and commons:

> The landed property of a ceorl (farmer) shall be fenced both winter and summer. If it is not, and if his neighbour's cattle come through an opening that he has left, he shall have no claim to such cattle, he must drive them out and suffer the damage . . . If free peasants have the task of fencing a common meadow or other land that is divided into strips, and if some have built their portions of the fence while others have not, and if their common acres or grasslands are eaten [by straying animals], then those responsible for the opening must go and pay compensation to the others, who have done their share of the fencing, for any damage that may have been suffered.

Fencing was ferocious, as good as today's barbed wire. The *Anglo-Saxon Chronicle* of AD 547 records Ida of

Northumbria constructing a hedge around his settlement at Bamburgh, impossible to penetrate.

Early monastic records show that cattle were regarded as almost exclusively draught animals, only to be eaten when worn out, and their value for milk and cheese negligible. (The 'white stuff' revolution came in the thirteenth century.)

Cattle were pre-eminent on the Anglo-Saxon farmstead, their value captured in law, which was dominated, as with Ine's legislation, by clauses, injunctions and penalties for the theft of beeves. The worth of cattle was embodied equally in language: in Old Saxon cattle were *fehu*, synonymous with the modern 'fee', the noun 'cattle' associative with our 'capital'. In a sort of sad proof of cattle's importance in *Englaland*, the Bayeux tapestry has a cameo in which gleeful Normans run off with a large white ox. That Norman robbery was cosmic justice; the Anglo-Saxons had white cattle because they had stolen them from the Britannic locals on their own arrival.

White-horned cattle in west England going on Wales were majestic figures in the landscape for a millennium or more, from Celtic time to Norman time. Rhys ap Gruffydd, who ruled the kingdom of Deheubarth in South Wales, is recorded as keeping white cattle. He died in 1197.

~

The history of land settlement is illumined by nomenclature. Place names are testimony to the activities of the Saxons. Their inroads into the forest are recorded for all time in such words as hurst, holt, hey and field – all are suffixes indicating clearings or places where livestock pastured. Then there is 'meadow', from the Old English verb *mawan*, to mow. So at Woodston there is Tining Meadow, Blackland Meadow, Crundall Meadow.

Surely, but slowly, the Anglo-Saxons in Britain came to understand grassland management, and the growing of grass for winter fodder. The very first legal evidence for grassland management is a grant by King Hlothere of Kent, dated 679, of an estate with meadows (*Laton prata*) and pastures (*pascua*).

~

The Saxons relocated the Woodston farmstead from on the linden ridge down to the flat eastern base of the hummock, where Woodston Manor (to give Woodston Farm its grandest title) is now situated. The Saxons liked their livestock close by – you can see it in the names and cluster of minute field-shapes around Woodston: Calves Close, Lucerne Patch, and bits of land so small they are nameless to us now. With land close by and permanent, naming was unnecessary effort. One could just point and say, 'There.' If naming was required, it erred towards the practical; one field at Woodston is still Near Lower Field. Farming is prose, not poetry.

For the basic necessities, the nameless farmers of Woodston were self-sufficient. They were freemen who ate what they grew, grew what they ate. Clothing came off the back of the livestock, roofing and bedding from the fields. Their long hall was wooden, sunk in foundations (and no squalid hovel), all of which befitted a freeman, holder of a 'hide', roughly 120 acres, or the land necessary to support a middling family, the area of land ploughable in a year. What Bede called *terra unius familiae*.

Weasels were kept as mousers, though the day of the cat was coming. (By Elizabeth I's time the agriculturalist Thomas Tusser regarded the cat as an essential item of farm equipment.) In winter, some of the stock may well have been brought into the hall to share the human abode, man and beast separated by a wooden screen. As in roundhouses, the animals' body heat provided the central heating.

The noise from the animals would have been a Babel at Woodston, this typical English farm. The Anglo-Saxons kept pretty much every domestic beast on the livestock list. Aside from the favoured cattle there were sheep (unimproved from Celtic time, only more numerous, as many as 7,500,000 in England); geese (a fowl on the up, with a boy to supervise them, the 'gooseherd'); the occasional goat (little favoured by the Saxons, but they lingered longest in the west of England and Wales); horses for the farmer to ride; and an awful lot of pigs, though these were generally kept on wood pasture, unless farrowing. Indeed, woodland became valued in the number of pigs it would support; it is a motif running through every vellum page of the Domesday Book.

Pigs being what pigs are, the swineherd at Woodston had a thankless task if he sought to prevent their straying. They doubtless interbred with wild boars coming down from the Wyre Forest to the north, and anyone who has tried to keep amorous porcines apart will feel his futility. (One of our Large Black boars – Large Black being a traditional breed as well as a descriptor – on failing to surmount the brick wall of the rented Georgian kitchen garden where I was running a few sows – again, why dig yourself if you have pigs to do it for you? – eventually decided to bulldoze said wall. He weighed in at about 340 kg.)

Anyway, the lingering wooded areas of Anglo-Saxon Woodston retained importance – they were managed for fodder, coppice, timber. In woodlands there were rights of pannage, of cutting green boughs (valuable for cattle). But hunting game such as the wild boar was a diminishing part of everyday life. In the excavated middens of the Anglo-Saxons, the bones of wild animals make up only a small percentage of the total bone waste created by butchering.

Domestic bones excavated from Cassington, a pagan Saxon village near Oxford, proffer a window into the diet of the time:

Ox 43 per cent

Pig 29 per cent

Sheep 12 per cent

Horse 8 per cent

Dog 8 per cent

Wild game barely figured, though even the keenest student of bones would be pushed to differentiate between the wild and domestic pigs of the ninth century AD, the both of them shark-toothed, hirsute and slow-maturing.

The reasons for the dominance of domestic meat over game in the daily meal of the Anglo-Saxon is plain. Hunting is haphazard, even for the best of hunters.

~

The farmer at Woodston did not live in an egalitarian world; Anglo-Saxon society divided into a simple two-tier hierarchy, eorl and ceorl. Eorls were the elite distinguished by birth, wealth and, for men, office; ceorls were all other free men who were farmers in one way or the other. However, a good deal of farmwork was done by slaves, the descendants of British natives. The Domesday returns from Welsh border counties point to the enslavement of tribesmen won in battle until Christianity came.

Initially, Woodston was one free scattered farmstead among many, but as time went by the diverse settlements of the first Anglo-Saxon occupation concentrated in new estates. Nucleated settlements began to replace dispersed farms, and the land of these settlements began to be sub-divided among those who had a share in it. That seems, on the available evidence, to be the context in which communally shared open fields evolved.

Woodston came under the control of Eardwulf of Earduleston (now Eardiston), then Eorl Wiferd, themselves subjects of the Kings of Mercia. Kings expected rents, delivered as food to the court. Ine at the end of the seventh century laid down the customary food rent expected from every ten hides (1,000 to 1,200 acres) as: 10 vats of honey, 3,000 loaves, 12 ambers of Welsh ale (an amber being half a mitta or 32 gallons), 30 ambers of clear ale, 2 full-grown cows or 10 wether sheep, 5 salmon, 20 lbs of fodder and 100 eels.

The Saxon farmer had other mouths to feed, aside from his family, his lord and his monarch; the Christian church brought Bede and the Luttrell Psalter, but it also brought a parson and his church to almost every village, and the parson and his church needed to be sustained.

The Saxon lord's contribution to the church was the endowment of the parish priest with glebe land, and his people had to provide voluntary gifts towards the parson's upkeep. This act of Christian charity eventually transmuted into the legally enforceable tithe. Thus, King Ethelred decreed the tithe-owner should receive 'the tithe of young live stock at Pentecost and of the fruits of the earth at the feast of All Saints'.

Tithes at Woodston for the upkeep of St Michael's, Lindridge, were collected into the Georgian and Victorian eras; the Vicar of Lindridge's account of 1796 itemizes 5 shillings received from William Mound of 'Woodson'; the tithe plan of 1840 itemizes the 'tithe-able' sections as Lucerne Patch, Rick Yard, Damson Orchard, Upper Bow Croft Orchard, Homestead and Calves Close.

Then there were the 'plough alms' (effectively another Church tax) the Christianized peasant farmer was expected to donate; these grew excessively from a penny for every team yoked to the plough to the 10 rams, 400 loaves, 40 dishes, 13 hens and 260 eggs which the men of Warmington had to render to the Abbot of Peterborough in 1086.

The wealth of *Englaland* flowed to the coffers of the Church in other ways. Christianized Saxon lords fell into the habit of rewarding bishop or abbot for his services (or ensuring salvation of their own souls) by the gift of estates. By the time the Domesday Inquest was undertaken in 1086 about a quarter of England belonged to the church.

While the peasant might render lip service to the new God, and trudge from Woodston to the Christian church on the hill, his heart still worshipped the old Teuton gods: Tew, the god of war; Thunor, the thunderer, and Frig, who multiplied the peasant's seed, in bed and in the field. (The Christian calendar revived the feast of Frig as the Harvest Festival: this was in 1843, at the instigating hands of Reverend Robert Hawker of Morwenstow in Cornwall.)

~

A short discourse on corn dollies. 'Dolly' is probably a corruption of the Greek *edidolon*, meaning 'apparition'. Long, long before the coming of Christianity to Europe, it was believed that the spirit of corn lived among the crop, and she took refuge in the last sheaf to be harvested. To give the corn spirit a refuge, these final straws were woven by the reapers into the likeness of a woman or into a geometrical cage. The effigy or 'dolly' containing the corn spirit would be carried home to the farmstead and hung on the wall the winter long. I have one, given me by a Cheshire farmer who wanted, in his old age, to remind the world of the old ways. Craft. Connection with the land. Cultivation of heritage cereal species.

~

Eorl Wiferd was among the Saxon nobles overcome with Christian kindness or the desire to secure a place in Heaven. Together with his wife Alta he granted Woodston and the whole of Lindridge to the Christian church of St Peter in Worcester in AD 796, in the time of King Offa of Mercia. ('Mercia' comes from an Anglo-Saxon word meaning boundary, also the derivation of the term The Marches.) The land charter, a conveyance drawn up in proper legal form, came complete with a perambulation, whereby the boundaries of the land were circumnavigated and the significant features noted down in writing.

The charter describing Lindridge declares the land as comprising 15 hides, though 'hide' here is not an exact measurement of area (120 acres) but a device for calculating taxes. The overall acreage is in excess of 6,000 acres.

The parish boundaries described in the charter are exactly the same as those on the Ordnance Survey map of today. The charter's southern boundary is the Teme at Woodston, where my mother and her four sisters swam, and where they fished for 'tiddlers' with jam jars and bait of white bread.

~

The Anglo-Saxon era was a time of birds. By opening up more of the wildwood to farming, the Anglo-Saxons increased the species of bird in Britain. Thanks to them, we have the kestrel – the people's falcon, be you lord of the manor or northern mining boy, across our skyscape.

~

All the length of the soft, white-skied morning, a kestrel has kept me company while I have been on the fencing. (In traditional farming, one does not do a job, one is 'on lambing', 'on haying', etc; I have my grandfather's fencing pliers, a very tangible inheritance.) Usually the kestrel has been as a fluttering mote – like something trapped on the retina – but sometimes it is so close I can make out the droopy moustache that gives *Falco tinnunculus* such a rakish appearance. If birds are to be anthropomorphized, the kestrel is the bounder who drives an open-top Morgan sports car.

Once, the kestrel stooped so near I saw its obsidian talons pierce a rodent's back, heard the latter's death-shriek. A cry from the past, millennia-old.

Our kestrels are doing well; they like our unkempt, bleach-tipped grassland because it is running with voles. Also, I have put out a couple of 'hoppers' for the supplementary feeding of the red leg partridge, the clown bird whose whirry, little-leg clockwork running brings a smile, always. Some of the spilled seed makes meals for mendicant mice, which in Nature's food chain makes food for falcons.

The kestrel is the quintessential farmland falcon, probably the greatest raptorial beneficiary of the open, 'quilt pattern' landscape our native agriculturalists axed from the wildwood – hence the kestrel's local name of 'field hawk'. Our kestrels assiduously avoid the scrub, the woods, the forest, always seeking and searching meadow and arable expanse. If, as the Provisional Wing of rewilders wish, our national tree cover is expanded from 13 per cent to 40 per cent on the wolfy, lynxy Continental model, where will our kestrels go?

Conservationists are too often aware of the wonder of woodland but ignorant of the importance of farmland.

I should have paid more attention to the kestrel. A kestrel always faces the wind when hovering – its tail spread, its body frenetically hunched and hugging. (The Anglo-Saxon for kestrel is, well, very Anglo-Saxon: 'wind fucker'.) A kestrel is as reliable a wind indicator as the brass cockerel weathervane on the Norman church steeple, the gaudy orange wind-sock at the airfield.

At some point, while I was pontificating, the kestrel pivoted to north. Lost in the anthems of my own opinions, I failed to register the aerial signal. And got drenched.

The Anglo-Saxons bequeathed us our love of birds. One Old English calendar for 1061 has 364 days devoted to saints. The remaining day? The eleventh of February, for which the entry reads, 'At this time, the birds begin to sing.' It was the only thing worth noting that equalled the godly activities of the Christian exemplars.

The pattern of the rural year moved from the bird-song and flowers of spring through summer's work of harvesting to autumnal stock-taking and into the chilling challenge of winter. Every year was different but the rotation of the seasons was consistent – a consistency strengthened by a calendar of proverbial lore, annual rituals, festivities.

Such was the picturesque land and satisfactory way of life evolved by the farmers at Woodston, and England generally, in the last years of the Anglo-Saxons. Water-mills, granaries and ovens were erected to cope with – and flaunt – the fat of the land. As arable farming grew, sheep and cattle came under closer management and lived longer lives, yielding more wool, dairy goods, and traction power.

It was a place of plenty, a temperate Arcadia. It was exactly this attractiveness that made the country once again worth invading.

~

The arrival on the English scene of William the Conqueror substituted Norman landlords for the old Saxon thegns, but farming methods and duties and the rights of men who farmed the land stayed much the same. The Normans had nothing new to offer in the way of agricultural techniques.

*The fat of the land: Woodston's hop-kilns,
'the size of a castle', c. 1950.*

CHAPTER IV

THE MANOR

The Domesday survey of Lindridge – Woodston becomes the 'Manor of Wodeston' – Open-field farming – The free warren – The market and the vicarage – The ploughman – A day in the life of medieval Woodston – Black Death and the decline of the feudal system

WOODSTON IS NOT MENTIONED by name in the Domesday survey of 1086 but is included in the 15 hides of land at Knighton and Eardiston held by the prior and convent of Worcester, which could support 'A priest, 15 villagers and 10 smallholders with 15 ploughs; a further 3 ploughs would be possible. 17 slaves.'[1]

The farmer at Woodston was one of the said 'villagers', a tenant of the church. At some unknown point early in the eleventh century Lindridge had been lost to the monks, Woodston with it, but William the Conqueror restored it to Bishop Wulfstan, who gave it to Thomas the Prior of St Mary's, Worcester. The Conqueror, alias William the Bastard, was elaborately considerate towards religious institutions; indeed, part of his claim for the throne of England had been that God was on his side.

Despite the changes of ownership, life at Woodston rolled languorously on, green and quaint, remarkably the same as before the Norman deluge. Crested peewits laid their eggs in spring among the sheeny furrows, the sweat of the August harvest ran down the back beneath the woollen shirt. Woodston was a small world, the farmer tied to its acres, from early youth to his death. Every tree and hedge, every nest and pond and woodland were as familiar to him as the veins and broken nails of the suntanned back of his hands.

It is the temptation of historians to seek out change; if you are a farmer you desire stability. History is an urban construct; farm life, even in the twenty-first century, still lives largely by the seasons, the weather, the immutable organic.

Also, history tends to the schematic.

As any child who has read *1066 and All That* at skool knows, the Normans initiated feudalism, which classically is a rural pyramid consisting of the lord of the manor (apex), who held his fenced-off land in demesne and not sublet to tenants, and which was worked for him by the subject classes; the middle of the pyramid consisted of the 'villein' (villager), a peasant occupier subject to the lord; next rank down was the cottager, who held a cottage and plot of adjacent land, usually about 5 acres, by virtue of a labour service. At the bottom of the heap was the bordar, who rendered menial service for his hovel and kitchen garden, which he held completely at the lord's pleasure. Officially, the Normans dispensed with outright slavery, though a bordar might quibble the truth, and the *servi*, serfs, certainly would, tied as they were to the demesne and working for keep.

In Norman society everybody was a tenant of somebody, including the King, who was a tenant of God Himself. The Norman class system became enshrined enduringly in the language of food and farming: where the Anglo-Saxon locals kept cows, the Norman lords ate beef (from the French *boeuf*); where the English farmed the sheep, the Normans ate the mutton (*mouton*).

Obligation in the Norman system, however, flowed both ways. The lord of the manor offered his subjects protection from bandits, bullies, foreign invaders, wolves – all and anything disruptive of home, hearth, the working of the land. William Langland, who set *The Vision of Piers Plowman* nearly within sight of Woodston, on the Malvern slopes, described the process thus:

> *All my life will I labour for love of thee*
> *if thou wilt keep my church and me*
> *from the waster and the wicked that would destroy us.*

Woodston had benefit of clerical masters, and also benefit of the protective Norman gauntlet. The farm was nearly within sight and sound of the Norman motte-and-bailey castle at Burford, built to defend and to control the crossing on the Teme that allowed communication and trade between Worcester and Ludlow. The Domesday Book noted the castle's owner as Richard le Scrope, Sheriff of Shropshire, who was tasked with securing the Welsh borders. His son, Osbern Fitz Richard, maintained his father's military manner.

So: Woodston was spared both Edric the Wild's 1068–70 savage revolt against the Normans, and the brutal Norman counter-harrying. North of Burford, William laid entire districts 'waste'.

~

In theory, so the Norman model goes, the indentured subjects of the manorial lord, whatever the rank given them, strove to produce food using the 'open field' system, with the peasants ploughing up and down long lengths with teams of communal oxen. Each peasant's 'strip' was separated from its neighbour by a ridge and double furrow – a pattern cast on the fields which can still be seen in places, especially in winter when vegetation is paltry and slumped, or from the air. Hitler's Luftwaffe, in its reconnaissance missions over England, provided an absolute bonanza of photographs of 'ridge-and-furrow' sites to interest the student of such things.

At Woodston, like much of the west of England, the theoretical Norman geometry failed to operate. There was no 'nucleated village' whereby all the villagers trooped out to two or three open fields, to do the same thing every day. Lindridge at its eastern end was at best a hamlet of a handful of houses. What happened at Woodston was messy (and rather English), a making-things-up-as-one-goes-along. Some fields were on the open system, some enclosed with hedges, especially pasture land and orchards; the stock would also have been put out to grazing on the common land of Frith Hill, some run on the remaining woodland, itself enclosed. (More of that later.)

Woodston in 1200 would have been absolutely median England in terms of land use: 35 per cent arable, 15 per cent woodland and wood pasture, 4 per cent meadow, 30 per cent pasture, the rest marshy bits. The coda is that 'woodland' would already have included Woodston's orchards. In the reign of Henry III, the county of Worcestershire was celebrated for the cultivation of fruit, hard and soft, including apples, pears, cherries, plums, raspberries, walnuts and chestnuts. (The county remained so into my childhood; my uncle Tom grew the blackberries for Ribena over at Kyre.) Some of the orchards at Woodston, such as Carters, Pool, Croft, were planted on the linden ridge; in other words, the high ground was re-treed.

History goes backwards, as well as forwards.

Woodston was a self-contained unit, wherein the farmer, a super-villein, must, in his best moments, when the grain was in, the calf born, his son or daughter on the long reins of the ox-plough, have felt like a little lord himself. Our Saxon free farmer and his wife have not disappeared. Their feet are planted firmly in the soil as they toil behind plough and swing the mattock – immortalized in the poignant picture of Piers the Ploughman and his wife at work in the cold field while their babies on the headland cry. Neither are their heads so bowed they fail to notice the opportunity of adding a few new acres to the holding, or profit from selling direct to the customer at the local markets. Woodston was on the up, out of obscurity.

~

The year 1135 hardly figures in the history books. No kings were made, or queens demised. A scribbled line in the Worcester Cartulary, however, notes that John Paschal of 'Wodesintun' granted his lands to the prior and monks of St Mary 'in his pious charity' and for 'the benefit of their souls'. He owed rents at Wodesintun on half a virgate, about 15 acres.

A century later another owner of Woodston (ownership in Norman England being by leasehold), Alured of Penhull, gave the lands to the cathedral priory of Worcester, because he had become indebted to Jews, and placed in prison in danger of forfeiting his life. Due to the good service of the shaven-headed monks of the priory of St Mary his debts were paid, in return for certain of his lands; the priory also undertook to provide Alured with a 'crannock' (an ancient Celtic unit, roughly two bushels) of grain, half wheat and half 'siligo' (soft wheat), every six weeks during his life, and a rent of half a mark at Michaelmas during the lifetime of his mother, and 10s. at the same feast every year after her death. Alured was also to retain a house and croft belonging to the dower of his mother, to inhabit during his life. (Another ancient family of Lindridge, the Lowes, were also saved by the priory from indebtedness.)

No longer are the farmers of Woodston nameless. Finally, the farmers of Woodston are named – John Paschal and Alured of Penhull. No one is anything without their name inscribed somewhere, on a tomb in a quiet country graveyard, in a parish record.

John Paschal and Alured of Penhull are also subject examples of the divisive, separating processes of Norman feudalism. They were both small farmers, hit by the rise in inflation in the twelfth century. So they went to the money lenders, then the wall. There was always someone to take their place at Woodston, because farming is a good dream, and a bad habit.

~

Little if anything about the Norman feudal system encouraged better farming, though it did value dung. Sheep fed on pasture during the day were folded (penned) on the arable land of the lord at night. It was said that dung of sheep was more valuable than wool or meat.

The Norman system, in agriculture as in politics, was exploitation, exploitation, exploitation.

Thus, Woodston had more land under the iron plough than ever before, the team of eight beef beasts able to drag a plough with a mouldboard through mirey, heavy soil – and thus irrigate it, with the furrows fabricating mini-dykes to allow the run-off of water. Moreover, land not under the plough – the woods, marsh, pasture – was more widely grazed.

A similar picture obtained across England. The country's population rose in line with productivity, from *circa* two million in 1086 to *circa* six million in 1347, making it one of Europe's most settled places. For those with a stake and a share in the system – the lords, and a few of the richer freeholders and peasants – these were, concurrent with the relaxing of Norman feudalism, good centuries, times of opportunity and profit. The farmers of Woodston were among the elite, the 15 per cent of landed peasants who held a full virgate of 30 acres, since they also held bits of land in neighbouring Dodenhill and Newnham.

~

The real growth of Woodston, however, came not from catering to towns and population increase but at the expense of the last unclaimed wildwood; by the thirteenth century the expansive wood that gave 'Wodeston' its name had all but disappeared.

Wolfless woodland had more value. A shepherd could walk unafeared, the axeman chop untroubled. Trees were

income, capital, faggots, kindling, charcoal, farm implements (from plough to cart, hoe to hurdle), pannage for
pigs, browse for cattle, sheep and goats, and the space
between the trees grassland grazing. (Today's 'silvo-
culture', 'agroforestry', is an old practice revived.) Other
uses of local Norman woodland included leaved branches,
especially of holly, for winter fodder, oak bark for tanning
leather in Ludlow, honey (a sweetener and the main ingre-
dient in mead), while beeswax was used to make fine
candles. Some honey was gathered from the nests of wild
swarms, and some from purpose-built hives, or 'bee stalls'.
Further potential income came from the capture and sale
of sparrowhawks and other birds of prey to falconers.
Farmers, or at least farmers who wish to survive, have
always diversified.

~

Stand at Woodston today and close your eyes to the post-
medieval buildings – especially the industrial-sized hop
kilns transformed into middle-class apartments – and the
lie of the green, gently swelling land would be near
identical to 1299. The soundscape similar too, because
by 1300 the howl of the wolf had vanished. The last
wolves in medieval England were confined to the west,
and the north. Edward I, in 1281, had employed Peter
Corbet 'to take and destroy all the wolves he can find
in . . . Gloucestershire, Worcestershire, Herefordshire,
Shropshire, and Staffordshire'. The campaign was

successful, and is commemorated by an iron wolf's head on a contemporary door at Abbey Dore.

The wild swine were extinct a century earlier. The last recorded free-living swine are those ordered killed in the Forest of Dean in 1260. The wild pig was a noble beast. The Normans loved the *chasse*, and, having extinguished the native stock of 'venery' – the animals worth the hunt – introduced the fallow deer and the oriental pheasant. The first clear mention of the pheasant comes on the menu of Waltham Abbey, in *circa* 1170, when it was ordered for their diet by the King; by Tudor times it was worth pound for pound the same as swan.

~

It was still a world of wood. The wildwood may have disappeared from Woodston, but trees themselves retained importance; the one stone building in the vicinity was the Norman church, twenty years under construction, on the adjacent, commanding linden ridge. A perch for the church.

A mature oak was the most valuable of trees, in cold cash terms, due to its value in building; Woodston's single-storey farmhouse of three rooms (central hall/kitchen, service room, parlour, all with rushes on the floor) was oaken-framed, its walls wattle and daub, the roof thatched straw from the field; wheaten and barley straw was longer then, made of golden hollow tubes to drink through (hence the 'drinking straw'), a winter fodder and bedding for livestock that alchemically captured the sunshine of summer.

If there was no longer 'wildwood' at Woodston — the views to the east and south were largely clear for a thousand or more yards, save for the scraggling alders and willows along the brook — there remained islands of green woods west and north, their remains today being Vicarage Wood, Crundall Coppice, Coney Green Coppice.

~

That 'coney' in Coney Green Coppice tells a story.

King Henry III granted to the monkish overlords of St Mary's and Lindridge fiefdom a 'free warren', an elastic term, but in this case meaning a rabbit ('coney') farm, hardly difficult to locate here, the clue being the field names Upper Coney Green, Coney Green Coppice, Lower Coney Green.

Rabbits had come to Britain with the Romans, as can be seen from the bones in their rubbish pits. With the recall of the Legions and the slow decay in the way of life that the Romans instituted, the rabbit disappears from the archaeological record. They were then reintroduced by the Normans from Spain in the twelfth century, who farmed them in flat-topped structures of earth called pillow mounds. Around these were wattle fences, and men to guard them. The guardians were there not to prevent Flopsy, Mopsy and Cottontail from escaping but to stop locals nabbing them for the pot. Like deer, rabbits were reared exclusively by lords of manors.

The introduced rabbit was delicate and needed cherishing in our climate. At first it could not dig its burrows in English soil, so these were made for it by drilling with augurs. Domestic rabbit-keeping persisted in Britain until the nineteenth century; 2.3 million rabbit carcases were sold in eight markets in the 1872–3 winter alone. The fur was used in clothing; loose hair, glued with shellac, made felt hats.

The rabbit, however, proved impossible to confine, and once outside the warren, bred proverbially, thus reducing its own worth. In the thirteenth century the market price of the rabbit carcase averaged 3½d plus a further penny for the skin. It was a luxury. Two centuries later the price had fallen to 2¼d. By the eighteenth century it was food for the poor. In 1953, such was the national disapproval of rabbits eating their way through Mr McGregor's garden and Farmer Brown's bean crop that the South American myxomatosis virus was introduced, which promptly killed 99 per cent of England's rabbits. Essentially, they went blind, then bled to death.

Rabbit redux. They are all over the linden ridge today, piles of their pellets on their lookout stations, former molehills. My grandfather hated rabbits, not for the damage they did (though seven adult rabbits munch through as much grass as one ewe), but, having endured multiple rabbit dishes as a boy, he was 'sick to the back teeth' with them on the plate. He preferred hare, which he hunted over the fields and hopyards of Woodston with a .410 rifle. Shot hare was hung in the cold pantry till its flesh started to blacken. Then he skinned it, and casseroled it with onions and carrots grown in the garden.

My mother and her four sisters, when they were children at Woodston, tended to avoid the cold pantry, with its hanging, rotting, falling-apart hares.

~

Initially, labour services on monastic estates such as Woodston tended to be relatively heavy.[2] But by the thirteenth century labour services on priory manors for the holder of a full virgate were as little as six days ploughing, eight days reaping, and some carriage services. However, it should not be assumed that the farmer of Woodston carried out such labours for the lord himself. Men like him in this neck of England paid money in lieu. Or had his own men do the duty for him.

He is already a yeoman farmer.

For the farmer at Woodston, there are distinct advantages to having monks as the lord above. Aside from the warren next door, which he may well supervise (and, if not, surely avails himself of, either by outright theft or a nod-and-wink favour with the warrener), the monks had enticed Henry III into granting them a market on Wednesdays at Lindridge.[3] In other words, the Woodston farmer has a venue on his own doorstep to sell his produce, and for his own wife to sell her baskets of onions and butter and her goat's cheeses and sometimes some wool, though most of the wool – which goes as far as Flanders – is taken by buyers from London. The farm's flock is mostly wethers, castrated males, because these give more wool than ewes.

In case anyone thinks rural life boring in medieval times, the same market attracted jugglers, thieves, minstrels, hawkers, and all the farming neighbours for miles around so one could lament with them the weather, the high price of anything bought, the low price of anything sold. The form of farming language alters; the content is timeless.

~

Only clergy and nobles were fat in the Middle Ages; everyone else worked themselves thin.

This hot spring week of April I have worked the farm in the style, the manner, of the peasant farmer of Woodston *circa* the Year of Our Lord 1300. (Authenticity was aided by our apple trees being in blossom, like those of Woodston's orchards, where cider apple trees were doubtless introduced in the medieval era.)

I have hoed by hand, scythed by hand, dug by hand, picked peas by hand, and even persuaded my horse, Zeb, to drag a single-furrow iron for a couple of lengths. I did this in the white sun and in the black thunderstorm. I had no oxen as such to 'curry' (groom) but did brush down the cows every morning.

Some beans I planted with a dibber, a bean per hole. Wheat from last year was flailed with two wooden staves, tied together by a leather thong. I plucked a goose – geese being a novelty growing in popularity on the medieval farm, as were ducks – and used its fat for cooking, including the bird itself.

I went to church; I got up with the lark, went to bed with the first call of the nightingale in the evening. The first cuckoo sounded on the third of the month, so I knew it was the last day to sow barley, and I did so by walking up and down the ploughed field 'broadcasting' the seed by hand, my boots clogging with dirt. That night, by candlelight, I read the 'round' about the migrant bird known as the 'Cuckoo Song':[4]

Sing, cuccu, nu. Sing, cuccu.
Sing, cuccu. Sing, cuccu, nu.

Sumer is i-cumen in –
Lhude sing, cuccu!
Groweth sed and bloweth med
And springth the wude nu.
Sing, cuccu!

Awe bleteth after lomb,
Lhouth after calve cu,
Bulluc sterteth, bucke verteth –
Murie sing, cuccu!

Cuccu, cuccu.
Wel singes thu, cuccu.
Ne swik thu naver nu!

I 'dagged' sheep (tidied up their backsides with shears).
I collected the hens' eggs. Chopped wood, twice a day,
every day for the fire . . .

Peasant work is brutal. One is a slave to the work, whether
it is for yourself or your lord. I, 'John of Wodeson', did
this, too, on a medieval diet. The drinking was easy; the
medieval peasant supped exclusively the alcoholic brews
of mead, cider and ale since water was potentially dis-
eased. Breakfast was 'white meat', meaning goat or
sheep's cheese, with bread dipped in sheep's milk. My
main food for lunch and dinner was 'pottage', meaning a
stew of meat, vegetables (peas, beans, leek, garlic and carrots) and herbs (sage, onion, thyme, savoury), the plants
grown in an authentic medieval kitchen garden.

The Medievals possessed an entirely responsible carnivorism: they wasted nothing of a slaughtered animal, and
ate it from nose or beak to tail. (Compare, contrast with the
pickiness of contemporary meat-eaters.) The flesh in my
pottage was as much offal as 'prime cuts' – it was liver, testicles, intestines, lungs, sheep's neck. The pottage was served
up on a daily bread plate, a 'trencher', made of millet.

I tell you this. Everything in the pottage was enjoy-
able, from the textured variety of the differing meats to

the flavouring of the herbs to the way the stew sank into the bread – it was something soft in the hardness of life.

My English medieval diet was about 3,000 kcal per day, and twelve hours' physical activity was necessary to prevent weight gain – which was, more or less, the length of my peasant working day.

Put another way: I had to work twelve hours a day to grow the carbohydrate-rich foods to enable me to work the twelve hours to grow the food.

~

Also on my medieval diet sheet was frumenty (aka furmenty, furmity), a thick wheat porridge. The renowned medieval cookbook *Forme of Cury*, 1390, advises this recipe:

> FOR TO MAKE FURMENTY. Nym clene Wete and bray it in a morter wel that the holys gon al of and seyt yt til it breste and nym yt up, and lat it kele and nym fayre fresch broth and swete mylk of Almandys or swete mylk of kyne and temper yt al, and nym the yolkys of eyryn. Boyle it a lityl and set yt adoun and messe yt forthe wyth fat venyson and fresh moton.

This produces a sort of meat-broth porridge. Any grains can be used, cracked lightly in a pestle and mortar. 'Seyt' is an old word for simmer; allowing the dish to stand lets

the grain swell and soften. The almond milk was for flavouring, the eggs for thickening; a pinch of saffron added
sulphuric, startling yellow. Saffron exoticizes.

Frumenty – the various alternative spellings are all
based on the Latin *frumentum*, meaning grain – stayed in
the cook's repertoire until well into the nineteenth century, mostly as a spicy, fruity sweet pudding. It was
rum-spiced firmity at the local fair that led Henchard
into selling his wife in Thomas Hardy's *The Mayor of
Casterbridge*.

~

At Woodston, as well as the monks of the priory, there
was the church next door, and religion's power over daily
life deepened. In the beginning of the Norman epoch,
Lindridge church was seemingly served by a priest on the
prior's estate at Knighton and Eardiston. By 1205 Giles
Bishop of Hereford confirmed to the monks a pension of
40s. from the church of Lindridge, and in the following
year an agreement was made between the parson of Lindridge and the prior, by which the prior was to receive
yearly 10s. from the church and the parson was to have
all the tithes of Moor and Newnham, the manors adjacent to Lindridge, except tithes of hay. A vicarage for
Lindridge was ordained in 1310, with the vicar to have a
court with a garden and dovecot.

The godly were expected to attend church, and if you
were the farmer next door there were no absences that

the parson's keen eyes, from his high point, could fail to spot. So the year was marked by inescapable services and rituals, conducted in unintelligible Latin, held in a stone nave painted with apocryphal murals to put the fear of God into one. Ever clever, the Christian church adapted itself to the pattern of rural life, and incorporated the previous pagan agricultural rites and dates. The church and farming went together in the Middle Ages as surely as did the ox and cart.

The liturgical year started with Plough Sunday, the first Sunday after Epiphany, which had originally been a sort of Dark Age trick-or-treat known as Plough Monday, where ploughmen danced through the street, asking for bread, cheese and ale, dragging a gaudily dressed plough behind them; if they were turned away by a householder, they ploughed a vengeful furrow or two in front of the abode. This was truly a ceremony of propitiation, a relic of a rite as old as the plough itself.

In Ancient Greece, Demeter, goddess of grain and agriculture, had been placated with an offering of the first fruits at a feast called the Procrosia, 'Before the Ploughing'. Christianized as Plough Sunday, the new ceremony saw a ceremonial plough brought into church and placed in front of the altar of the Ploughmen's Guild, which was lit with tapers of rush or wax paid for by the local husbandmen, in order to ensure success for their labours throughout the year.

Candlemas, 2 February, the date of the Purification of

the Virgin Mary, marked the end of sexual abstinence practised since Advent: it was unwise in agricultural economies to have heavily pregnant women or newborn babies in the late summer months when there was work to be done at harvest, thus the Church, leading family planning, advocated celibacy in December, January and in Lent.

There were pancakes on Shrove Tuesday, to eat up the larder before the fasting period of Lent. Particular dates in the liturgical calendar – the feasts of major or local saints – prohibited work, among them, Easter, Ascension, Pentecost, Corpus Christi and Christmas. The first of August, the 'lammas' or loaf-mass of St Peter ad Vincula, was the blessing of the bread made from new corn. The celebration of the first new loaf was also marked by the removal of fences that had been erected on lammas meadows at Candlemas – common land that had been kept free of stock, growing hay that could then be cut and stored for winter. In the west of England, Lammas Day was also the traditional date to start haymaking, of no small importance to the birds of the ground, who had by then largely finished their nesting cycle.

Then there was St Martin's Day, 11 November, also known as Martinmas or Martlemas, when stock was slaughtered for Christmas, hence the proverbial wisdom on the inescapability of fate: 'His Martinmas will come, as it does to every hog.' The tradition of November slaughter certainly pre-dated Christianity; the Old

English name for the month was *blotmonap*, blood month. Those with a weather eye might observe that 'if the wind is in the south west at Martinmas, it keeps there till after Candlemas'.

So, the liturgical cycle became an aide-memoire for the agricultural one. Candlemas equalled erection of fences, loaf-mass their removal. It would be a mistake to think the medieval church mirthless. Latin Communion may have tested the patience of even the most ardent Christian farmer, but a sense of festivity was built into medieval church ritual. Carols were originally tunes for dance steps for specific saints' days.

If anything, the sense of festival, sacred and profane, grew in popularity throughout the latter Middle Ages. England determined to be 'Merrie England'. Why? According to the historian of the labouring classes, E.P. Thompson:

> Many weeks of heavy labour and scanty diet were compensated for by the expectation (or reminiscence) of these occasions, when food and drink were abundant, courtship and every kind of social intercourse flourished, and the hardship of life was forgotten . . . These occasions were, in an important sense, what men and women lived for.

Almost all gone now, the communal agricultural festivals, save for Harvest Festival, a Victorian invention to resurrect Merrie England via Christian-rural. Introduced by Reverend Hawker in 1834, it captured the imagination of the country-obsessed nation, perhaps stirring memories of old Lammas-tide services.

I do, however, remember Mrs Cole's cottage in Withington, East Herefordshire, around 1970 decorated inside with boughs of hawthorn on May Day. Foliage had a sacred importance for rural folk, and bringing spring greenery to the home was a declaration of the common good, ensuring that human and natural worlds would exist in harmony throughout the year; it was a form of correspondence with the land. My grandparents absolutely decked out their house with bines (stems) of hop in September so the house was filled with its bittersweet scent.

My grandmother also knew some of the words of an old folk song about 'maying', with each verse ending 'Fa la la'.

When I was young, I would ask her how old she was, and she would rejoinder, 'As old as my nose, but older than my teeth.' On her death in 1996, I learned she was born in 1910. But she could have been born in 1510, because the song I think she was singing was Thomas Morley's 'English Pastoral' for the sixteenth century, which starts:

> *Now is the month of Maying,*
> *When merry lads are playing. Fa la la.*

Each with his bonny lasse,
upon the greeny grasse. Fa la la.

Fa la la.

~

In north Worcestershire and south Shropshire, goats probably played an important part in the peasant economy of the early Middle Ages. Goats can subsist on marginal land and have a remarkable ability to convert underwood and rough and moorland grazing into meat, milk and cheese.

One of the enigmas of English culinary and countryside history is why and when the English lost their taste – unlike Continental Europe – for the milk products of goat and sheep.

Medieval ewes produced 7 to 12 gallons of milk during a lactation, and according to Walter of Henley, 20 lactating ewes gave enough milk to make 4 pints of butter and 250 round flat cheeses a week.

Sheep's cheese in England died out in the nineteenth century, goat's cheese earlier still. So today, we buy Roquefort from France, and Parmigiano from Italy.

~

Although most commercial production of wool now comes from descendants of Spanish merino sheep (introduced by the Beni-Merines, a tribe of Arabic Moors in the twelfth century AD), the sheep that made Britain's medieval wealth were home-grown native stock, and of two principal types: a longwool (from the Cotswolds), and the aforementioned shortwool Ryeland, grown on the unimproved rye-grass pasture around Leominster Benedictine Priory (counter-intuitively, sparseness of sward makes wool fibres finer). Certainly, the good monks of Leominster passed on their ovine-rearing secrets to other priories locally, including St Mary's of Worcester and its Lindridge holdings. 'Lemster Ore' it was called, the Ryeland's wool, for its ability to be parlayed into gold.

One Italian merchant recorded the prices for English wool between 1317 and 1321: Lemster Ore, the golden fleece, was worth 28 marks per sack of 363 lbs. (A mark was two-thirds of a pound.) At today's prices, 28 marks would be worth £3,000. When I've sheared Ryelands I've received about £180 a sack. Then again, the modern Ryeland has been crossed numerous times in unwise ways to improve the carcase, and the rule of 'improvement' is more meat, lower quality wool.

Anyway, sheep were the making of Woodston's wealth, as they were of England's wealth, which is why to this day the seat of the Lord High Chancellor in the House of Lords is a large square bag of wool called the 'woolsack' – a reminder of the principal source of English wealth in the Middle Ages.

～

Recently, I read the 'how to' manual of a fourteenth-century French shepherd, Jean de Brie. Shepherding was a career he took pride in, noting that biblical characters such as Moses and David were also shepherds, and that 'numberless people take their living, food, and support, for the most part, from the profit and gain of sheep'.

What has changed since medieval times is human respect for sheep. It has reversed. Here is de Brie on bringing up ovines:

> First of all, the lambs, young and tender, should be treated kindly and without violence and should not be struck or corrected with switches, sticks or whips nor any other kind of beating that could hurt or bruise them, for they would fall off and become thin and weak. Rather one should lead them gently and kindly by leadership and correction.

On shearing:

> In May the weather is fair and calm and not yet too hot. Everything on earth is in full flower, for then she has put on her beautiful gown, adorned with many lovely little flowers of diverse colours, in woods and meadows – it is then that the pastures are filled with beautiful, tender plants. In May it is the custom to shear the wool from rams, ewes, yearlings, and lambs, since the wool is ready then. It is also

more appropriate and greatly profitable to shear the sheep then than at any other time, as much for the season's moderate heat as for ease in pasturing.

Of himself as shepherd:

From experience which is the greatest teacher, he learned through great application the theory, practice, science, and manner of feeding, tending and managing woolbearing animals, and the natural law shown and taught to all animals, not only those who reason, but to all other beasts that are born and live, in the air, on earth, and in the sea.

There is no reason to suppose that the peasant farmer of Woodston in the 1300s was any less considerate towards his flock than de Brie.

The peasant farmer knows that those who abuse Nature pay a heavy price. Such as those who practise factory farming of animals. Sheep included.

~

Fish were important in the medieval diet, both nutritionally and increasingly as a means of mitigating the religious prohibition of meat-eating during days and seasons of fast. By the thirteenth century, many of the clergy and well-to-do laity avoided meat by eating fish every Friday and Saturday, for the season of Lent, and on the vigils of the main feasts. Additionally many households also observed Wednesday as a fish day.

The mill at neighbouring Newnham – of great antiquity and lasting until the 1850s – like most water-mills, probably had eel and fish traps set in its water channels (some Domesday mills owed their rents in eels). At some unspecified time in the Tudor and Stuart era, Woodston gained its own mill, quite likely at the same time that it gained its first pond, the millpond doubling up as fishpond.

After the Roman de-camping in AD 410, no fishponds existed in Britain until the members of the Norman secular aristocracy constructed them to enhance their status. But the real boom in fishponds came in the fourteenth century, and was partly due to more rigorous observance of the Christian calendar. They were a significant investment; the only people who could afford them were the aristocracy and monasteries.

Medieval fishponds had a steady flow of water through them – such as the nameless south-flowing stream at Woodston – to prevent the water becoming stagnant. The water was also relatively shallow because the fish needed to be caught easily. By the same token, the banks were kept free of growth so a net could be dragged across the pond. Bream was by far the most popular freshwater fish on the table in the thirteenth century, and it retained favour for nearly four hundred years among the well-to-do and the discerning, until its replacement by the carp, a native of the Danube. Carp grow rapidly, and reach large weights. Commercial production of carp enabled the middle classes and minor gentry to ape the plates of their betters. In *The Canterbury Tales*, Chaucer's Franklin has a fishpond.

The need or the want of a fishpond energized the Georgian and even the Victorian gentry, and not just for fish as Friday lunch, but for the sport of fishing; the two fishermen's pools upstream of the millpond at Woodston were created by the Adams family in the century between 1839 and 1926 (as that very Victorian exercise in scientific geography, mapping, shows us). The ponds – Mill Pool, Swallow and Kingfisher – remain today placid and reedy, and permits can be bought for their fishing; they are stocked with bream, carp, roach and crucian carp. (Farmers really do diversify.) The last time I saw the ponds, I put up a heron, which cranked into the air, fierce-eyed, and disgusted by my intruding. On a previous visit a kingfisher, fired from the imagination of Calliope, Muse of Poetry, bolted across the cluster of pools, its back the same cobalt blue as the sky. A mirror match.

Easy, is it not, to castigate us humans for the wrongs we do Nature? But sometimes we benefit it; almost all the ponds in the English countryside were dug by farmers, and the Victorian century was the golden time of the British pond. The number of ponds over twenty feet across in England and Wales in 1880 was about 800,000, or fourteen ponds per square mile.

Farmers dug ponds for the fishing, for the milling, but also as watering holes for the livestock and the horses. Farmers gave the heron, the dragonfly, the moorhen their killing ground. Their home.

~

Fish were, and are, money, and fishponds became a favourite target for poachers. Peasant discontent often expressed itself by assaults on fishponds, which were seen as bastions of privilege. The pond of the past was the unlikely site of class war.

My grandfather escorted, at midnight, West Midland hop-pickers away from the Woodston ponds, where they had been casting an illegal line; then again, as a boy, he had been asked to vacate the ponds at Burley Gate brick quarry, his catgut line flung out into the depths for eels.

~

The three ponds of Woodston are an elegant, liquid reminder that the best of farming is the multiple use of the same space: the ponds were providers of direct protein for humans (fish), as well quenching the thirst of the stock (thus indirectly providing protein for humans).

I was pondering this yesterday, as I took the public footpath from Broombank down through Woodston's orchards (much of Woodston is accessible by public footpath, avoiding that unpleasant feeling of trespass; the footpaths, pounded into the landscape, are the old access tracks of the Saxons, and before them the Prehistorics), spring in the air, a green woodpecker swimming wavily between the trees.

I grew up with such orchards, the grass between the trees eaten by sheep, then geese, then chickens; the trees in autumn providing fruit, not just for humans, but for the stock, with the pigs clearing up the rotten apples and the windfalls. Multiple cropping. Quite productive, the old ways of farming, if you think about it.

The old ways remain etched in my mind, and are memorialized in field names. One of the orchards closest to Woodston Manor farmhouse is titled 'Pig Run or Damson Orchard'.

~

Worcestershire was among the very first places where cider orchards are recorded as being planted, as early as the thirteenth century. When my grandfather Joe Amos was farm manager at Woodston in the 1940s and 1950s, he kept pigs in the orchard – Gloucester Old Spots, whose black marks are folklorically believed to be caused by falling apples. In his lifetime pork and beef were still smoked up the farmhouse chimney, partly to save money, partly for the taste. The wood for the smoking was apple.

In his childhood people still 'wassailed', went around the apple orchards on Twelfth Night, beating the trees with sticks and sprinkling cider around the roots, singing a song to urge them to 'bear many a plum, and many a pear'. All to ensure a good crop in the year ahead.

Wassail, from the Middle English 'waes hael' – good health – by way of Old Norse, is a pre-Christian festival, even a relic of the sacrifice made to Pomona, the Roman goddess of fruits.

~

The medieval inhabitants of Woodston lived by the flowers, because some, like the daisy, open and close as regularly as clockwork. Flowers were, and are, time-indicators. Poppy chimes bright and early in the morning, Nottingham catch-fly lies lazily abed till seven in the evening.

Carl Linnaeus celebratedly noted the botanical rhythm of flora, and how species differed from each other in their timekeeping, when out sauntering in the summery Swedish countryside. Enchanted with the reliability of 'clock flowers', or 'aequintocales', Linnaeus devised a full-scale *Horologium Florae* in 1745 for the botanical garden of Uppsala. This arranged the timely plants in the shape of a clock face, dividing the bed into twelve segments. A living clock. The wedding of Horticulture to Horology.

To Linnaeus go the bouquets, but 'clock flowers' were already well known before he gazed at goatsbeard and pondered proliferous pinks. The English poet Andrew Marvell described an elementary floral dial in 1678:

How well the skilful gardener drew
Of flow'rs and herbs this dial new;
Where from above the milder sun
Does through a fragrant zodiac run;
And, as it works, th' industrious bee
Computes its time as well as we.
How could such sweet and wholesome hours
Be reckoned but with herbs and flow'rs?

'Day's Eye' itself is a modernizing of the ninth-century Anglo-Saxon *dæges ēage*. Daisy. In this island we have been telling time by the flora of wayside and mede, wood and water's edge since the Neolithic dawn.

The people of the past were like flowers, their lives regulated by the length of natural light; they were 'photoperiodic' themselves. Their days began at bindweed's opening, they slipped off to bed at cowslip's close. Time was a different concept then. More approximate. Not regulated to the atomic second.

When I was a child, my grandmother's venerable friend Mrs Cole, who lived in a thatched cottage inherited from Old Mother Hubbard, believed that God decorated our countryside with clock flowers to aid us poor humans in our work and in our meetings. Scientists have removed the divine from the exegesis of flora's circadian rhythm, positing instead a theorem that it is all about competition and pollination. If every flower were open 24/7, profaners say, the bustling bees and hither-thither hoverflies would be overwhelmed by choice, so some flowers would be unvisited.

Whether designed or evolved, our British wildflowers remain as reliable as any time-keeper planted in a Scandinavian botanical garden.

To use a floral dial is to live in the old ways, to find a more natural rhythm, to revolt against the tyranny of the timepiece and the iPhone.

A BRITISH WILDFLOWER CLOCK

MORNING
   IV   Goatsbeard opens
    V   Poppy, Chicory open
   VI   Bindweed, Common Nipplewort open
  VII   Coltsfoot, Buttercup open
 VIII   Cowslip, Scarlet Pimpernel, White Water Lily,
            Rough Dandelion open
   IX   Lesser Celandine, Daisy, Proliferous Pink open
    X   Sorrel opens
   XI   Common Sow Thistle, Star-of-Bethlehem open
  XII   Goatsbeard, Common Nipplewort close

AFTERNOON
    I   Proliferous Pink closes
   II   Scarlet Pimpernel closes
  III   Rough Dandelion closes
   IV   Coltsfoot, Sorrel, Chicory close
    V   Water Lily closes

  VI Poppy closes
 VII Nottingham Catchfly opens
VIII Evening Primrose opens
  IX Cowslip closes

Time for bed, says Flora.

~

The first outbreak of plague known as the Black Death swept across England in 1348–49. It seems to have travelled across the south in bubonic form during the summer months of 1348, before mutating into the even more apocalyptic pneumonic form with the onset of winter. It hit London in September 1348, and spread into East Anglia all along the coast early during the new year. By February 1349, it was ravaging Worcestershire.

There are no records of who at Woodston lived or survived. If any did. It was very rare for just one person to die in a house. Usually, husband, wife, children and servants all went the same way, the way of death. Any family would be lucky to have a survival rate of 50 per cent. On the Bishop of Worcester's estates, death rates ranged from 19 per cent of manorial tenants at Hanbury to 80 per cent at Aston. The tell-tale of contracting the disease was a 'buboe', a swelling. The Welsh poet Jeuan Gethin wrote:

Woe is me of the shilling [swelling] in the arm-pit; it
is seething, terrible, wherever it may come, a head

that gives pain and causes a loud cry, a burden carried under the arms, a painful angry knob, a white lump. It is of the form of an apple, like the head of an onion, a small boil that spares no-one. Great is its seething, like a burning cinder, a grievous thing of an ashy colour. It is an ugly eruption that comes with unseemly haste. It is a grievous ornament that breaks out in a rash. The early ornaments of black death.

Jeuan Gethin died in March or April 1349.

Nor was 1350 the end of it. Plague recurred in 1361–64, 1368, 1371, 1373–75, 1390, 1405 and continued through the fifteenth century. There is some evidence the plague became age-specific as it matured, predominantly killing the young. By the 1370s, the population of England had been halved and it was not recovering.

A portrait of the consequences of the Black Death in the countryside is given by Henry Knighton, canon at the abbey of St Mary of the Meadows, Leicester:

In the same year there was a great murrain of sheep everywhere in the kingdom, so that in one place in a single pasture more than 5,000 sheep died; and they putrefied so that neither bird nor beast would touch them. Everything was low in price because of the fear of death, for very few people took any care of riches or property of any kind. A man could have a horse that had been worth 40s for half a mark (6s 8d), a fat ox for

4s, a cow for 12d, a heifer for 6d, a fat wether for 4d, a sheep for 3d, a lamb for 2d, a large pig for 5d; a stone of wool (24 lbs) was worth 9d. Sheep and cattle ran at large through the fields and among the crops, and there was none to drive them off or herd them; for lack of care they perished in ditches and hedges in incalculable numbers throughout all districts, and none knew what to do. For there was no memory of death so stern and cruel since the time of Vortigern, King of the Britons, in whose day, as Bede testifies, the living did not suffice to bury the dead.

In the following autumn a reaper was not to be had for a lower wage than 8d, with his meals; a mower for not less than 10d, with meals. Wherefore many crops wasted in the fields for lack of harvesters. But in the year of the pestilence, as has been said above, there was so great an abundance of every type of grain that almost no one cared for it.

Meanwhile the King sent proclamations into all the counties that reapers and other labourers should not take more than they had been accustomed to take, under the penalty appointed by statute. But the labourers were so lifted up and obstinate that they would not listen to the King's command, but if anyone wished to have them he had to give them what they wanted, and either lose his fruit and crops, or satisfy the lofty and covetous wishes of the workmen ... After the aforesaid pestilence, many buildings, great and small, fell into ruins in every city, borough, and village for lack of inhabitants, likewise many villages and hamlets became desolate, not a house being left in them, all having died who dwelt there; and it was probable that many such villages would never be inhabited.

The farming world was turned upside down.

In place of the serf, bound to the soil and lord by inalienable duties, there arose the landless labourer, fugitive from the manor which was now powerless to restrain him, and offering his services for cash to whomever would employ him. Instead of peasants occupying their strip fields by virtue of service rendered to their lord, there came into being a race of tenant farmers paying a cash rent for their land, and free from hereditary servile duties.

Many manors succumbed, and archaeologists enjoy the quest for 'deserted medieval villages' (there is one such DMV at Little Hereford, ten miles from Woodston on the A4546, a sequence of ghostly mounds and troughs in the earth, which become all the more depressing with understanding; by the time I finished my second degree in History, I kept my eyes on the road when passing). Also, much of the land of England, in an overcrowded countryside, had been worked out; thin soil, good only for pasture, had been put under the plough, and continuous cropping over many centuries with too little fertilizer had gradually impoverished the soil to the point at which diseases and pests killed off the crop as surely as the bubonic plague killed off humans.

~

With landowners facing both a labour shortage and impoverished earth, one solution was to have less acreage under the plough, and more put down to pasture for manuring livestock. Thus, the long trend towards increased arable farming in England was reversed.

More land for pasture meant more sheep in England.

*Joe and Marg Amos at Woodston, c.1938, with daughters
(left to right) Margaret, Daphne, Kathleen (mother of
John Lewis-Stempel).*

CHAPTER V

THE GOLDEN FLEECE

The Tudor and Stuart wool trade comes to Woodston,
where the super-profits are used by the Penells
of 'Woodson House' to elevate themselves to the
squirearchy – Ryeland sheep – The Civil War –
Puritanism in the countryside – Hops and new crops

IT WAS NEARLY TWO hundred years before Woodston
emerged from the post-Black Death void and into the
light of the record books once more. The 1522 journal of
William More, Prior of St Mary's, contains the entry:

> Item: Received of Roger Penhull [Penell] for ye
> rent of lyndrige £3 6s 8d.

What had happened to Woodston in the intervening two
centuries? The extraordinariness of the prior's entry is worthy of dissection. The records of mid-second millennium
England are incomplete, despite the assiduousness of pale-
faced monks. Time and cupboard-clearing tidiness have

done for many once valued notes. That is by the by. The fate of Woodston, as became the norm for farms after the plague, was to be doled out in whole or in parts to any tenant who would take it. All of them were once someone's dream. Their fate? To be swallowed by the neighbours.

Farms: they live, they die, they mutate.

But Woodston rose, phoenix-style, intact.

The Tudor rent for Woodston, Gloriana on the throne, was cash, plus a gift, on top of those obligations to the parson. The customary gifts from a tenant farmer to his landlord were described by *The Poesies* of George Gascoigne, 1575:

> And when the tenantes come to paie thier quarters rent,
> They bringe some fowle at Midsommer, a dish of Fish in Lent,
> At Christmasse a capon, at Mighelmasse a goose.

Goose was by now a patriotic dish. Elizabeth was said to have been feasting on goose on St Michael's Day 1558 when she was informed that the Spanish Armada had been defeated. The bird was traded in particularly large numbers at Michaelmas fairs as a winter Tudor necessity, the fat keeping out the cold when rubbed on the chest.

Wool, though, was the salvation of Woodston. Land, as I mentioned at the outset, can be interrogated, and I've walked Woodston in the light of plain midsummer's day, seeking and questioning. The evidence is there in the hedges. Up at the top of the farm, the hedges are five hundred years old plus; you can tell by the numbers of tree and shrub species in a thirty-yard stretch, a truth divined by the botanist Dr Max Hooper. Roughly, multiply the number of tree and shrub species in the stretch by a hundred and you have its birth century. The Hooper Rule works because a hedge acquires more species as it ages. The Woodston hedge species are hawthorn, blackthorn, hazel, holly, elm.

Through the fifteenth and sixteenth centuries, the landscape of Lindridge was gradually enclosed with wobbly-lined, square-ish, hedged fields, as local landowners built up their wealth through sheep. Arable cultivation declined steeply. These sheep were no longer the free-range, half-feral, half-goat animals of old, but ovines to be grown big and fat and, especially, woolly. That required genetic selection. That required control of the flock. That required fields with hedges, where DNA-good could be separated from DNA-bad, and the rams or 'tups' kept apart, so they did not indiscriminately impregnate every ewe. (Rams do so want to do that; one of our rams, half Suffolk half Shetland, a biblical progenitor of offspring, once climbed a wire stock-fence four feet high to pizzle every soft white pedigree Beulah next door.)

Woodston went over to sheep, at least two-thirds of it. That was England then.

~

In spite of the 'murrain' of the Middle Ages, the sheep population of Britain grew, and grew, and grew some more. In 1341 Parliament had granted Edward III 30,000 sacks of wool; assuming a fleece of about 1.5 lbs in weight and a content of 364 lbs wool per sack, this levy alone implies a population of 7,280,000 sheep of sheerable age. With the same rule of thumb, by the 1600s there were at least 12,000,000 sheep in Britain.

The ploughman was toppled. There was more money in fleece than turned soil. Exported abroad to Flanders, the fleece of sheep was truly golden. It provided England with the super-profits that financed exploration, stability, and the naval means for self-defence.

There were particular beneficiaries of the sheep trade, among them the Penell family of Lindridge, back from 'Penhull' penury, and very definitely on the Tudor-and-Stuart way up. On the back of sheep, the Penells took over Woodston as an independent farm, and made themselves gentry.

Good King Henry helped. The land that came on to the market when Henry VIII dissolved the monasteries in the 1530s – the ecclesiastics were the biggest landowners in Britain – was the making of many a farmer. The Penells leased land, then bought the freehold.

~

This, I think, is the place to introduce to the story of Woodston my mother's maternal family, the ap Harris.

According to the bard they paid to write their family tree – and no bard worth his medieval mead ever traced genealogical roots to riff-raff – the ap Harris arrived in west England about 1200, from Wales, where they were allegedly descended from a Viking king of Ireland and a Welsh princess. (Or maybe Norse princess and Welsh king; who knows.) Anyway, on the Welsh edge of Herefordshire the ap Harris established themselves as feudal stewards, doing royal business. They fought at Agincourt as men-at-arms along with their equally hardcase relatives the Vaughans and the Gams, Sir Roger Vaughan and Sir Daffyd Gam dying to protect the king (or so the Tudor propagandist Shakespeare conjures in *Henry V*). The ap Harris' star rose and rose. By the time of Queen Elizabeth they had become respectable Anglo-Welsh gentry, retitled and anglicized as Parry, and constituting something of a power at court. It helped that their 'coz' was William Cecil, Lord Burghley, the chief layman in the land.

It was Blanche Parry, Elizabeth's chief lady-in-waiting, who gifted the woollen hose Gloriana so enjoyed, and inspired the bucolic poet Michael Drayton to dithyramble:

> Where lives a man so dull, on Britain's farthest shore
>   To whom did never sound the name of Lemster Ore
>   That with the silkworm's web for smallness doth
> compare.

The wealth from sheep had a paradoxical effect on Woodston, in that it enabled the farm to grow sufficiently to become two units. By the late 1500s Woodston had split into Upper and Lower Woodson farms, both held by the Penell family.

~

A good sheep, the Ryeland, docile and round as a teddy bear, with a sweet carcase. I can confirm too that Ryelands, being heavy grazers (with little if any browsing instinct), improve the sward because they'll eat any type of grass, even rough 'rye', enabling rich and varied herbage to flourish on the ground. They were, along with Red Poll cattle, chief keepers of my traditional, flower-rich meadow up in the Black Mountains.

There are long lines of continuity in the countryside. Big circles too. I farm the sheep my Parry ancestors twenty times back bred. The same sheep breed made Woodston's fortune in the sixteenth century.

~

Predominantly pastoral Woodston never went through the apocalyptic enclosures of the Georgian era; it had never possessed much open field and was enclosed early.

The sheep and the other animals played their part in the building of England's pleasant land. Enclosure, landscaping and the teeth-work of the beasts achieved an idealized pastoral scene, the 'quilt' of green fields, stitched together by hedges; the landscape of Woodston.

The animals, in the history of our countryside, never get their due, do they? Woodston and England were ideal for sheep, sheep were ideal for England and Woodston. Sheep were and are extremely effective at extracting energy from natural vegetation. The eighteenth-century Agricultural Revolution of Jethro Tull is taught in schools; what the textbooks omit is that the first farmyard revolution was in medieval times, when sheep were 'folded' on the arable part of farms to manure them. A sheep is a walking muck-spreader. A living, organic machine for fertilizing ground.

The humble wool of the back of Ryeland sheep was the source of England's wealth, stability, power. Democracy. (As Trotsky once pointed out, as soon as you have bread queues, you need a policeman to keep the people in order. Then a policeman, to keep an eye on the policeman . . .)

Poor sheep. Overlooked in our history, and now in the twenty-first century accused of abetting climate change through their 'gaseous emissions', and destroying the landscape with their hooves and mouths. According to rewilders our land is 'sheep-wrecked' when it could be covered bounteously with trees. These new forests enveloping our meadows, alas, will be a sad surprise to the meadow pipit I saw at Woodston the other day.

Also, if we get rid of all the sheep, we get rid of wool – warm, natural, sustainable. Is that not a sort of madness?

~

A short essay on sheep shit. As the Victorian sheep expert William Youatt expressed it so eloquently:

Sheep's dung is valuable for manure, and for some other purposes. It has been supposed, and probably with truth, that it contributes more to the improvement of the land than does the dung of cattle. It contains a greater proportion of animal matter, and that crot densed into a smaller compass; and it falls upon the ground in a form and manner more likely to be trodden into and incorporated with it, than the dung of cattle. Hence arose the system of folding sheep on the arable part of a farm in many districts in the midland and southern parts of England. The sheep were penned on a small space of ground, and the pens being daily shifted, a considerable quantity of land was ultimately manured.

At Woodston, sheep were fed on pasture during the Tudor day, folded on the arable fields at Tudor night. Win. Win.

~

A real come-back-from-the-grave family, the Penells. These Tudor Penells are, of course, the descendants of Alured of Penhull, whom we met in the twelfth century.

They worked Woodston for four centuries – clearly some fragment of the family escaped the buboes of the Black Death – raising the place, raising themselves. In 1523, one Willyam Penell paid the prior a whopping 104s 2d and 37s 6d 'for ye rent of Lyndriche'.

Willyam Penell was accumulating land from other farms, generally to the north (including on the far side of what is now the A456), and east towards Dodenhill, including land once owned by the priory, such as Monk Meadow. When he died in 1544 he left, under his will, bequests to Hereford Cathedral, Lindridge, Mamble, Stockton, Eastham and Knighton churches and to Elizabeth Tondyn, the remainder to his son, Walter.

Skip forward to 1660, and the Lindridge parish register has the entry 'Edward Penell of Woodston, Esq. buried'. The 'esquire' is no politeness, it is a distinct social rank. The Pennells had entered the ranks of the gentry.

The unacknowledged truth of farming is that it has always been a ladder for social mobility. Nothing reflects the growing prosperity of Tudor and Elizabethan husbandmen more than the comfortable homes they erected in the late sixteenth and early seventeenth centuries. So many lovely houses of rural England were constructed then that it has been called the age of 'the great rebuilding'.

So, no surprise: the Penells built themselves houses befitting their status and aspirations. Upper Woodston House is currently Grade II listed by Historic England:

Farmhouse, now house. Early C17, refronted mid-C18 with mid-C19 and mid-C20 alterations. Timber-framed with brick infill on sandstone rubble base, brick refacing and additions, plain tiled roofs. L-plan; main part of three framed bays aligned east/west; large external rubble chimney with two star-shaped brick shafts to rear of west bay; also large external chimney to rear east bay enclosed by two-bay rear wing. Two storeys, attic and cellar; dentilled eaves cornice. Framing: exposed at west gable end and rear; seven rows of panels from sill to wall-plate at gable end, five rows of panels at rear; long straight braces at lower corners of both storeys. Roof structure appears to have been altered, probably when building was refronted; west end collar and tie-beam truss has two long struts to collar and subsidiary strut to lower rail at each side. South front elevation: three bays; all windows have gauged flat heads and stone sills; outer bays have 15-pane sashes; central first floor window is a 10-pane sash; central entrance has a moulded flat canopy, a moulded architrave, and half-glazed door with panelled reveals. Attic lights in gable ends. Interior not inspected. Main entrance in west side elevation of rear wing.

It's a far cry from the villein farmers of Wodeston, who would have sat down on a wooden bench to eat off plain pottery placed on a trestle table, and slept on a straw pallet.

Robert Furse, a late-sixteenth-century Devon yeoman, made an appropriate comment: 'Although our progenytors and foreforefathers wer at the begynnynge but plene annd sympell men and women and of small possessyon and hablyte, yt have there by lytell and lytell so run ther corse that we are com to myche more possessyones and credett and reputasyon than anye of them h[a]dde.'

~

By 1600 Woodston (Lower Woodston as it was now termed) took some working, given there were no labour-saving agricultural machines for its 500 acres; indeed, since the Anglo-Saxon age farm technology had barely progressed, the only change in locomotive power being the switch from ox to horse, the latter being quicker at ploughing and carting. At Woodston the change-over was marked by equines receiving dedicated grazing, hence Horse Pasture field. The horses, big and brown, were surprisingly delicate, and required much cossetting.

No longer did the ploughman and carter stare at the back end of a cow; they stared at the hindquarters of a horse. One day this view too would be lost, replaced by the view over the tractor bonnet. There are many lost views down on England's farms.

~

Many hands were required on the Tudor and Stuart farm, all the year round, and not just for the set seasonal tasks of shearing and harvesting. Feudalism with its ties and obligations was all but dead, so the Penells, like other farmers, hired labour at hiring fairs or 'mops' held quarterly. The most important of the hiring fairs was the one held at Michaelmas. Under the First Statute of Labourers (1351), a worker could go to be hired at the nearest market town on the day after Michaelmas Day – that is on 30 September – which was when pay rates were set. The other main quarterly days for hiring were Lady Day, May Day and Ascension Day. Thomas Hardy describes the hiring scene in *Far from the Madding Crowd*, 1874; it would have been no different to 1574:

> At one end of the street stood from two to three hundred blithe and hearty labourers waiting upon Chance – all men of the stamp to whom labour suggests nothing worse than a wrestle with gravitation, and pleasure nothing better than a renunciation of the same. Among these, carters and waggoner were distinguished by having a piece of whip-cord twisted round their hats; thatchers wore fragments of woven

straw; shepherds held their sheep crooks in their hands, and thus the situation required was known to hirers at a glance.

A 'good man', meaning an industrious and sober worker, could haggle his pay rate. Accommodation tended to be the hayloft, although long-term hands were already being 'tied' to a cottage. The Mill Cottage at Woodston was built for an employee in the late 1700s, though, as if acknowledging its spiritual ancestry, bore gabled dormer windows, an architectural feature from Tudor times.

It was Mill Cottage, enhanced, that my farm manager grandfather received with the job. Like many houses in Lindridge, it used to flood.

~

The major hiring fair for Temeside farms-on-the-Tudor-up was held at Sturbridge (now Stourbridge), twenty-five miles by road from Woodston, which came with the added attraction of being the prime place to sell farm goods and buy household necessities. Between the end of the sixteenth century and the middle of the eighteenth, this was the largest fair in England. Granted its original charter in 1211, it ran until 1933. At the height of its success and popularity it lasted for an extraordinary thirty-five days. The main trade was hops and sheep, although it expanded to include every conceivable commercial interest. The author Daniel Defoe described it thus:

It is kept in a large corn-field, near Casterton, extend-
ing from the Side of the River Cam, towards the
Road, for about half a mile Square . . . It is impossible to describe all the parts and Circumstances of
this Fair exactly; the Shops are placed in Rows like
Streets, whereof one is called Cheapside; and here, as
in several other Streets, are all sorts of Trades, who
sell by retale, and who come principally from London with their Goods; scarce any Trades are omitted,
Goldsmiths, Toyshops, Brasiers, Mercers, Drapers,
Pewterers, China-Warehouses, and, in a Word all
Trades that can be named in London; with Coffee-
houses, Taverns, Brandy-Shops, and Eating-Houses
innumerable, and all in Tents and Booths.

Defoe, somewhat ironically, omitted to mention the book-
sellers. The demands of agriculture meant that there
was a huge market in books offering advice for the farming year. These so-called almanacs were, by the later
seventeenth century, outselling every other publication –
including the Bible. Almanacs combined three varieties of
essential information: first, the almanac proper gave de-
tails of astronomical events, eclipses, conjunctions and
moveable feasts, based on the calculation of Easter; second, the 'kalendar' listed days of the week by date, month
and fixed festivals; and finally, the 'prognostication' was
the astrological forecast for the year. The moon was

believed to affect the planting of crops, so the lunar information contained in almanacs assisted farmers in their planting and in their slaughter around different phases of the moon. Rudolph Steiner's 'biodynamic' farming, a sort of ultra-organic agriculture, works by the moon's phases to this day.

~

As well as almanacs, the Tudor and Stuart farmer had the enjoyment of textbooks about farming. Two such books in particular in the mid-sixteenth century enjoyed a heedful audience, beyond anything that had gone before, one written by Fitzherbert, the other by Thomas Tusser. Fitzherbert – who was probably a Derbyshire landowner and High Court judge – brought out his *Boke of Husbandry* in 1523. It was the first printed agricultural handbook in England. He wrote from knowledge; he knew, for example, that if a ewe is well fed before she goes to the ram she will have more eggs to be fertilized. He also distinguished between three varieties of barley, and six of wheat. (The first proof that cereal crops were evolving into the hybrid forms of today.)

However, it was Thomas Tusser who set the benchmark in farming self-help books. This Essex farmer, born in 1524, published his first textbook in 1557, which was expanded into *Five Hundred Points of Good Husbandry*, probably the farming bestseller of all time.

Thomas wrote in doggerel, at once risible, at once impossible to forget. His advice on keeping the best cows was:

> *No storing of pasture with baggagely tit,*
> *With ragged, with aged, and evil at hit,*
> *Kept carren and barren be shifted away,*
> *For best is best, whatsoever ye pay.*

Every farmer likes a bit of autodidacticism on the parlour bookshelf. For my grandfather it was the blue-leather-jacketed *Country Gentleman's Estate Book*; for me it is Lady Eve Balfour's *The Living Soil*, published in 1943.

To be honest, they are much the same.

~

In Tudor England the farmer as we think of him appears fully formed: simultaneously industrious and just a little imperious.

The Tudor farmer had the framework of the enclosed farm in which he could improve his materials and his methods. Also, he had the new literacy, based upon the newly founded grammar schools, which enabled him to read what other farmers were doing.

Never an easy thing, education in rural society, since as often as not it offers a route away . . . and hankering for the life left behind. I offer, among many proofs, my aunt Daphne, who became a teacher, married a teacher, became a hotelier in Weymouth. Then, when her parents died, bought their house. Near Woodston. Where she herself grew up.

~

The earliest extant picture of Woodston and Lindridge embellishes a 16'9" x 13'9" tapestry map owned by the Bodleian Library of Oxford. The map has not been on public display since 1908, and, in avoiding the bleaching of daylight, retains its original brilliant yellows, reds, greens and blues.

It was commissioned *circa* 1590 by Ralph Sheldon, a rich landowner with tracts of countryside in Beoley, north Worcestershire, and Barcheston in south Warwickshire. Since he was a Roman Catholic, thus forbidden high office, Sheldon determined to make his successful status known to the world by constructing a grand mansion in the parish of Long Compton, which lies at the junction of four counties: Worcestershire, Warwickshire, Oxfordshire and Gloucestershire.

To adorn the four hall walls of his manor house, Sheldon decided on four tapestry maps, one for each of the neighbouring counties. (Tapestry, like the building of country piles, was conspicuous flaunting of wealth; Sheldon even shipped over Flemish weavers, the elite of the tapestry craft, for the job.)

For the design of the tapestries Sheldon derived his inspiration from two recently published books, both present in every gentleman's library. These were Christopher Saxton's book of county maps, published in 1579, and William Camden's *Britannia*, a travelogue published in 1586. Sheldon added roads, and accurate pictures of towns, villages and country houses.

Lindridge is shown with its defining hill ridge, wooded possibly with lime trees, along with the roads to Clows Top and to Mamble.

The tapestry's borders have Camden's verses embroidered into them, and this one surely describes Lindridge:

Here hills do lift their heads aloft, from whence sweet
*   springs do flow*
Whose moisture good, doth fertile make, the valleys
*   couched below*
Here goodly orchards are, in fruit which do abound
Thine eye would make thine heart rejoice, to see so
*   pleasant ground*

Sheldon's tapestry was one use for the wool of the Golden Fleece Age, for the rich, at least. For the poor, the enclosures for sheep provoked protests, particularly when peasants were dispossessed to make room. During Jack Kett's rising at Norwich in 1549, peasants slaughtered 20,000 sheep, a deeply symbolic act.

Theoretically, the Tudors loved the yeoman farmer, like Mister Penell of Woodston, who had a stake in the system. A conservative. But his sheep, despite giving the fleece off their backs to the national coffer, were deemed a white scourge. Tudor governments railed paranoically against sheep because shepherds were 'ill archers', never finding the time or inclination to practise with bow-and-arrow on the village green and thus defend the nation. Moralists such as Thomas More lamented the enclosure of fields around the country, writing movingly of the families which lost their cottages when their little farms were enclosed: 'the poore, selye, wretched soules, men, women, husbands, wives, fatherless children, widowedd, wofull mothers with their yonge babes . . . awaye their trudge . . . what else can they doo but steale and then be hanged?'

If the morality of enclosure is dubious, the economic results were beyond dispute. Cereal yields had fallen very low by the fourteenth century after hundreds of years of over-cropping and under-manuring. When the wool boom came, arable fields became pastures and the land had a much-needed rest from the blades of the plough.

Time was on the side of the farming squirearchy. Tudor Britain became a massive wool factory. What land did go to the plough saw rotation with grass (and thus a good dose of excrement from the rear end of livestock), and yield rose markedly in the sixteenth century. As a rule of thumb, farmers were harvesting at least double the medieval yield.

Part of the improvement in yields came from more

scientific and more varied organic fertilizing. Walter Blith's *English Improver Improved* (1652), which rivalled Tusser's *Five Hundred Pointes* in popularity, listed as soil improvers: 'Liming, Marling, Sanding, Earthing, Mudding, Snayle-codding, Mucking, Chalking, Pidgeons-Dunge, Hens-Dunge, Hogs-Dunge, Rags, Coarse Wooll, Pitch Markes and Tarry Stuffe, and almost anything that hath any Liquidness, Foulness, Saltnesse or good Moysture in it.'

The 'coarse wooll' is worthy of note. In the 1940s my grandfather used to spread 'shoddy', wool remnants from the carpet factories in Kidderminster, over the hopyards. By then, Britain's wool trade had been in centuries-long decline.

~

In 1570 Lindridge received its own Act of Parliament, which directly impacted on farming at Woodston. It was enacted to confirm the traditional rights of the copy-holders (tenants who were given a copy of their agreements in writing) of the parish. After Henry VIII's Dissolution of the monasteries in the 1540s, the Priory's landholdings in Lindridge had been given to the Dean and Chapter of Worcester Cathedral. The Dean and Chapter, however, denied the local inhabitants their traditional rights, as exercised under their previous owner, the Priory. It is Protestant-Tudor propaganda to depict monks as porky, selfish crooks; by the standards of the time, the Priory had been good landlords to Lind-ridge for four hundred years. (Witness their saving of Alured Penhull.)

Faced with the new, aggrandizing Dean, the people of Lindridge, including Woodston, took the Dean and Chapter, their owners, to court. Not just locally, but to the highest court in the land. The good country folk of Lindridge went to London, to Parliament. Not only was their case upheld but it was enshrined in law. The 1570 Act confirms their ancient rights, including to have pos-session of all woods, underwoods and buildings on their copyhold lands, and to have common pasture for their sheep and cattle. From then on the manor court books, which were kept by the Dean and Chapter as a record of business, always had a summary of the Act of Parliament of 1570 written in the front, as a reminder of who was who, and what was what.

~

Of course, the fates of English farming are akin to those of the Greeks, another agricultural people. After hubris comes nemesis. The Penells, like many in the west of England, supported the King in the Civil War, and paid the price. They were, if not outright Catholic, so High Church they were as good as. In the redbrick wall of Upper Woodston there is still a niche for a statue of the Virgin.

With the triumph of Cromwell's Commonwealth

in 1649, the Penells had their Woodston land confiscated for defending the cause of the Stuarts. They remain, however, immortalized by their epitaphs on the walls of Lindridge Church:

In memoriam WLLM PENELL de WOODSTON GENT qui in vxore ELIZABETH ROUDON obijt 31 Jan 1623. This stone that covers earth and claye, Longe in the earth uncovered laye, Man for'st it from the mothers wombe, And made thereof for man a tombe, Ane now it speaks and this does saye, The life of man is but a daye, The daye will pass the night must come, Then here poore man is all thy roome, The writer and the reader must, Like this good man be turned to duste, He liveth well and soe doth thou, Then feare not death when where or how It comes, And will end all grieff and paine, And make thee ever live againe.

Here lyeth the body of Edward PENELL of Woodston Esq. who departed the 5th day of August in the year of our Lord 1657. Here rests his earthly parts whose soul above views his bright makker face to face and prove pure joys . . . shal be full and perfect when these broken organs shal be peic'd again and reinformed. Reader before thou passe take this example a cleare looking-glass to dress thy soule by leave of Him to bee good in bad times who mayst live worse to see.

Woodston, after the Penells – a century of churn:

1675 From the Lindridge parish registers: John Mound of Woodston buried

1704 From the Victoria County History: Woodston Estate sold to Thomas Baker

1729 From the Lindridge parish registers: Edward Mound of Woodston buried

1745 From the Gloucester Record Office: Rev Thomas Baker sold 'copyhold lands in the manor of Lindridge, inc 1 messuage [house and associated land] and ½ yard called Calves Land with appurtenances in Woodson'

1746 From the Lindridge parish registers: Arthur Nott of Woodston buried

The land is eternal; its 'owners' transient.

~

As well as confiscating land, the Puritans of the 1650s sought to impose misery upon it. Lindridge church in Tudor and Stuart times had been vibrant with colour and High Church paraphernalia. The inventory of 1522 had itemized:

> Silver items, a chalice, patent and pix, and brass items, two crosses, two lamps, a pax, a candlestick and a censor. Of lead, were two cruets, and of pewter, twelve

little shells to stand under tapers, to set on the rood screen. There were two altar cloths and a cloth hanging before the high altar. A large collection of vestments included a violet velvet cope and vestments of velvet, of changeable colour, green Russell worsted and blue damask. There were various items of linen, including banners and streamers, and a collection of bells. In the steeple were three bells, there were also a lych bell [a corpse bell, the origin of 'death knell'] and 'eight little bells upon a wheel'.

The Puritans took away the vestments, the banners and the streamers, and made the place 'plain and godly'.

Under the Commonwealth of Oliver Cromwell, festivities and revelry were repressed, and real frost put on the life of Merrie England. Christmas was abolished, since there was no scriptural authority for it; in 1657 the diarist John Evelyn was arrested at gunpoint in church on Christmas Day for celebrating the Nativity. (Old Father Christmas became a rallying point for Royalists; after the Restoration, Christmas developed into a highly convivial affair, with gift-giving, pantomimes and carols.) Dancing around the village maypole was condemned by one Puritan apologist as 'stinckyng'. Apart from their theological objections to the bacchanalia of the maypole, Puritans disapproved of the wastefulness of spring festivities in the long century of poor harvests from the 1550s to the 1640s. The unpredictable weather which flattened crops could be read in biblical terms as an expression of divine displeasure at the corruption of the world and the fickleness of human faith, and as a reminder of the Fall. (The great floods at Woodston must have caused consternation; no wonder the suspiciously papist trappings were removed from the church.) By the late seventeenth century one English economist was complaining that each public holiday cost the country £50,000 in lost labour (c. £6 million today). Money: the root of all good for Protestantism and nascent agri-industrialism.

The regulation of festivity, sobriety, more work days. The new rule of the land was ceaseless labour, all year round: the Protestant work ethic in the field. Farmers and farmworkers became separated, and the response of the downtrodden was to politicize the calendar festivals that remained, and figures such as the Lord of Misrule came to preside over the revels. The characteristic style of English rural celebration – the costumes and the cross-dressing, the masquerade and the songs – became ritualized, mythologized and invested with layers of meaning.

Oak tree symbolism took off during the English Civil War, by virtue of Charles II's stay in the Boscobel Oak after losing the battle of Worcester in 1651. With the Restoration in 1660, the King's birthday of 29 May became Royal Oak Day, Oak Apple Day, Oak-ball Day, or Shick-shack Day, when a piece of oak (the 'shick-shack') was worn on the person or placed on the house. Oak

Apple Day was declared a public holiday by Parliament, and its official celebration continued until 1859. It was a major festival in Lindridge.

Another tradition which survived Puritanism was the blessing of the apples (as distinct from and complementary to the January blessing of the trees themselves, 'wassailing'). The Herefordshire antiquarian John Aubrey noted in *circa* 1750:

> on Midsommer-eve, they make fires in the fields in the waies: sc [scilicet, to wit] to Blesse the Apples . . . I doe giesse that this custome is derived from the gentiles, who did it in remembrance of ceres her running up and downe with Flambeaux in search of her daughter Properia, ravisht away by Pluto; and the people might thinke, that by this honour donne to ye Goddesse of husbandry, that their Corne, &c. might prosper the better.

Oak Apple Day and the Blessing of the Apples notwithstanding, communal celebration in the countryside was slowly dismantling.

~

The soundscape of Woodston changed during Tudor and Stuart times, for the worse. The avian aubade enjoyed by John Paschal would have been muter by the time the

Penells ended their tenure of Woodston. There were fewer birds, although farming practice was not responsible.

Vermin. The word comes down to us via Old French from Latin *vermis* (worm). King Henry VIII decided many wild birds and animals were as lowly as the worm, and required execution. Under the Preservation of Grain Act, passed in 1532 after a series of bad harvests, it became compulsory for every man, woman and child to kill as many creatures as possible that appeared on an official list of 'vermin'. And 90 per cent of Britain's 3.5 million population lived in the countryside, where the 'worms' were. Henry VIII put a bounty on each targeted creature, ranging from a penny for the head of a kite or a raven to 12d for a badger or a fox. These were considerable sums when the average agricultural wage was around 4d a day.

All parishes had to raise a levy to pay for the bounties, while communities that failed to kill enough animals were punished with fines. The inducements were successful; a million bounties were paid for hedgehog heads alone in the latter half of the seventeenth and first half of the eighteenth centuries. The hedgehog was subjected to wholesale persecution because of the erroneous belief that it sucked milk from the teats of recumbent cows at night. The act priced the head of a hedgehog at 4d – four times that of a polecat, wildcat, stoat or weasel. The Vermin Laws were strengthened by Elizabeth I, who put the price on the head of a wildcat at 1d. If the wildcat survived around

Woodston until the seventeenth century, it was now done for; almost 5,000 bounties were paid for wildcat heads in England and Wales in that hundred years.

Chuchwardens oversaw the Vermin Acts; in Lindridge, the churchwardens tended to be the occupiers of Woodston. Not for the last time, government policy would lead farmers down the path to silent fields, and their own ruin.

~

Introduced to England and Woodston Farm during the Tudor and Stuart era:

Leek – first mentioned in 1557.

Onion – in the reign of Elizabeth I.

Carrot – 1558.

Pea – in the reign of Henry VIII.

Turkey – introduced here from North America, possibly by Cabot in 1497. The agriculturalist Thomas Tusser, writing in 1573, refers to turkeys as being common.

Spinach – 1568.

Potato – generally ascribed to Sir Walter Raleigh in 1585 or 1586, but when the herbalist John Gerard was writing in 1597 it was still uncommon. The Royal Society appointed a committee to report on its potentialities in

1662, and by the end of the century it was being grown quite extensively in gardens. It was not until the late eighteenth century, however, with the growth of a teeming urban population demanding cheap and nutritious food that it became regarded as a farm crop. Transition to this status was most rapid in Ireland, where the population rose from 1,500,000 to 4,000,000 souls in the course of the eighteenth century. Having come to rely to a dangerous extent on a potato diet, the Irish were overwhelmed by an unprecedented disaster when the appearance of potato blight in 1845 and subsequent years caused the potato famine. The English retained their disregard for the potato, staunchly preferring wheat for their staple food, which was endorsed by the Lord's Prayer and had an almost miraculous genesis through the actions of yeast, being 'risen again'. The Georgian journalist and farmer William Cobbett maintained with his customary forthrightness that potatoes should be fed to pigs, not people, and that they were the cause of 'slovenliness, filth, misery and slavery'. Neglected as a field crop in the nineteenth century, the potato only came into its own during the Second World War. The government began actively promoting the economic advantages of potato cultivation to feed the people.

Sycamore – Native to central and eastern Europe and western Asia, it was probably introduced into Britain by 1500, in the Tudor period, and was first recorded in the wild in 1632 in Kent. Pollarded, it made good firewood (easy to saw, split with an axe, producing a hot flame). It achieved a footnote in British history when, under a sycamore at Tolpuddle in Dorset, England, six agricultural labourers formed an early trade union in 1834. Found to have breached the Incitement to Mutiny Act 1797, they were transported to Australia, and into trade union legend as 'The Tolpuddle Martyrs'.

Clover – Clover as a plant had been present in British grassland for millennia; these white field varieties were now complemented by the larger Dutch clovers. The virtues lie partly in its value as a highly productive and nutritious crop for grazing by stock or for conserving as hay; and partly in the way in which it improves the soil. Its long root systems penetrate and break up compacted subsoil; its leaves fix large amounts of nitrogen from the air, which are dispersed, via the root system, into the soil to feed succeeding crops.

Turnip – about 1550, beginning cautiously the revolution in farming whereby livestock was overwintered on 'roots'. Turnips have great agricultural virtues. They are best grown in widely spaced rows, or drills, and this makes it possible for men to hoe between them and kill the weeds. They are thus 'a cleaning crop'. They yield very heavy; a good modern cereal crop of field grain will produce about 2 tons of fodder per acre. Turnips will grow up to 50 tons per acre, plus the 'tops', the green leaves. Turnips can be 'fed off' in the field, or picked and stored. And in time, both turnips and clover were to become the cornerstones of the 'crop rotations'.

Hops – native to England, but not cultivated here until the sixteenth century. The first planting in the West Midlands has never been established, but it is likely to be prior to 1636, for in that year there was a reference to a field in Littleton, Worcestershire, which was named The Hopyard.

Woodston was not far behind.

CHAPTER VI

THE GOOD SEED ON
THE LAND

*The eighteenth-century Agricultural Revolution –
Enclosure – Jethro Tull's famous rotating cylinder seed
drill and Charles Townsend's four-crop rotation (wheat,
turnips, barley and clover) – Increased efficiency and
fertility at Woodston – Gentlemen farmers – The diseases
of the farmyard – The Napoleonic Wars*

THERE IS NO AVOIDING it: the expansion of Woodston
Farm included some enclosure of common pasture, of
the open fields. I'm thinking of the fenny, rough-grazing
east of the main farm, which became the two hedged-in
fields with the same name, Part of Stephens Moor. Also,
the flatland of East and West fields. If in west Worcestershire enclosure had proceeded piecemeal since the Black
Death, it nonetheless had a cumulative effect, the displac-
ing and dispossessing of the small farmer and the rural
worker. Enclosure at Lindridge continued into the nineteenth century.[1]

Before enclosure, the rural labourer could provide for his family by working for wages, but also by 'cropping' and running stock on the common land, the people's pasture. After enclosure – an early form of privatization – the labourer had no opportunity for small-scale farming. He became landless, status-less and likely confined to his parish by the Settlement Act, which prevented those receiving poor relief from moving to more generous parishes. Or he could migrate to a burgeoning manufacturing town, which, in the orbit of Lindridge, was the smoky metropolis of Birmingham.

Rural roots are hard to sever. When my mother was a girl at Woodston in the 1940s some of the seasonal hop-pickers were the descendants of Lindridge labourers gone to 'Brum' a century before. They came 'home' every September. When they departed their villages, they created little bits of countryside behind the new house in the city, the lawn standing in for the meadow, the veg patch for arable field.

~

The weeping pain of enclosure for the rural poor is written by their one true Logos, the Northamptonshire poet John Clare. In verses such as 'The Mores', Clare, son of a farm labourer, documented the effect of enclosure, granted to local landowners, on what had been open heath and moor available to all. Before the Acts of Enclosure:

Unbounded freedom ruled the wandering scene
Nor fence of ownership crept in between
To hide the prospect of the following eye
Its only bondage was the circling sky
One mighty flat undwarfed by bush and tree
Spread its faint shadow of immensity
And lost itself, which seemed to eke its bounds
In the blue mist the horizon's edge surrounds
Now this sweet vision of my boyish hours
Free as spring clouds and wild as summer flowers
Is faded all – a hope that blossomed free,
And hath been once, no more shall ever be
Inclosure came and trampled on the grave
Of labour's rights and left the poor a slave
And memory's pride ere want to wealth did bow
Is both the shadow and the substance now . . .

What enclosure accomplished was the destruction of the ancient English peasantry. The whole story is told in detail by J. L. and Barbara Hammond in *The Village Labourer*. From 1702 to 1762, 246 private enclosure Acts, covering about 400,000 acres, were passed by Parliament. From 1762 to 1801 some two thousand Acts dealt with over 3,000,000 acres, while from 1802 to 1844 another 2,500,000 acres were covered by just short of two thousand Acts. By 1914, 6.8 million acres of the countryside had been enclosed.

Enclosure swept away community traditions, such as

'gleaning' of wheatfields by parties of children and women, who went in after cutting to pick up the spare grain. Such traditions had helped community cohesion, reminded the squirearchy of their duties, of *noblesse oblige*. (Enclosure also gave birth to the image of the large farmer as a greedy and grasping man. Too often, the image had, and has, some substance.) Anxious that enclosure might cause unrest, Parliament resorted to brutal penalties for any transgression. Much of the notorious 'Black Act' of 1723 was designed to defend landowners' property, making it a capital offence, for instance, to 'break' or destroy fishponds. (Woodston had three of them, the Georgian farming squirearchy continuing the monks' taste for bream and carp.) After 1741, taking a single sheep could lead to the gallows. Murderers only accounted for about 10 per cent of the Georgian executed – the other 90 per cent were victims of this 'Bloody Code'.

The dispossessed took to drink. According to that old radical countryman William Cobbett:

> Go to an ale-house of an old enclosed country, and there you will see the origin of poverty and poor rates. For whom are they to be sober? For whom are they to save? (Such are their questions) For the Irish? If I am diligent, shall I have leave to build a cottage? If I am sober, shall I have land for a cow? If I am frugal shall I have half an acre of potatoes? You offer no

motives; you have nothing but a parish officer and a workhouse! – Bring me another pot –

By coincidence I read these lines of Cobbett before popping into tiny Tenbury Wells, to go to Tesco. I passed the Pembroke House (fifteenth-century inn), the King's Head, the Market Tavern, the Crow, the Vaults, the Ship Inn. If I'd carried on towards Woodston, I'd have passed the Bridge, the ex-Swan pub, the Peacock Inn (where my cousins Charles and Rob once earnestly tried to persuade me that drinking Fernet Branca was a prophylactic for a hangover), then the Talbot, a bit of a family gathering shrine.

God, if rural drinking is correlated to misery then Tenbury Wells must have been the saddest place on earth.

Which is odd, because I've always thought it the loveliest town in England.

~

The other effect of enclosure was on the environment. Marshes were drained, rivers canalized, and the fields squared and regimented. Intensive usage of the land was to the detriment of the snipe who dwelled in the mire, the glow worms who lit the heath, the corncrake who hid among the crops. According to Clare: 'birds and trees and flowers without a name / All sighed when lawless law's enclosure came . . .'

All that Clare loved was torn away by enclosure. With the destruction of his heartland went the deterioration of his own mind. Bouts of depression on top of poverty, dismay at enclosure, seven children to feed, a forced move from his childhood village of Helpston, felled him. He moved only three miles to Northborough, yet it was enough: 'I've some difficulties to leave the woods & heaths & favourite spots that have known me so long for the very molehills on the heath & the old trees in the hedges seem bidding me farewell.' Clare and his landscape were one; take away one, lose the other. (His poetry thereafter became preoccupied with nests.) At the age of thirty, he was put into an asylum near London. Four years later he walked out one July morning and in split old shoes made his way back home, eighty miles, sleeping in barns, his head towards the north to show himself the steering point in the morning, living on a chew of tobacco, grass, and a sup of ale when some kind stranger tossed him a penny.

When rural roots are severed, the consequences are hard to bear.

~

Enclosure also gave England its distinctive hedgerows. (By definition enclosure required a perimeter hedge.) Dudley Stamp in his book *Man and the Land* (1953) came up with a figure for the total length of the UK's hedges at 1,500,000 miles. The national hedgerow was a linear woodland, a wildlife reservoir, a hedgehog's dream. So, to a degree, enclosure offset Clare's nightmare of environmental destruction.

For the farmworker, the hedge was somewhere to sit under for lunch, out of the winter wind, the summer sun. Somewhere to puff the clay pipe, or, in my grandfather's day, a Woodbine cigarette; to draw on the jug of cider, or later the brown bottle of Bulmers. I have worked on and lived on six Herefordshire/Worcestershire farms; every single one of them has had, deep in the black base of their hedges, discarded cider bottles.

~

The most widespread hedgerow tree planted by the enclosure movement is the hawthorn, dense and robust and relatively stockproof. (Also known as the May bush, it provided the May Queen's crown.) Under the Enclosure Acts landowners sometimes insisted that their tenants planted elm at intervals along the hawthorn hedge; the object was to both yield timber and enhance the look of the landscape. The elm towered above the hedge bushes, a defining image of the English pastoral scene – until Dutch elm disease wasted it in the 1970s. Between 1970 and 1977 eleven million trees were destroyed, and the English hedge robbed of much of its splendour. As a child I watched the elms felled in our south Herefordshire village, one by one, always hoping that one special tree at least would survive, not turn yellow, its leaves wither, but it did not. The sound of the summer of 1976? The Sex Pistols singing 'Anarchy in the UK' and chainsaws, which my father suggested were remarkably similar.

The causative organism of Dutch elm disease is a fungus, *Ceratocystis ulmi*, which produces a substance that causes the cell walls of the elm to produce, against further fungal invasion, sticky swellings called tyloses; these block the vessels in the trunk of the tree that carry water to the leaves from the roots. The tree dies of thirst, the condition exacerbated by hot conditions, like the summer of '76.

The English elm, typical of Worcestershire, the Midlands and the Welsh Border, was particularly susceptible to the disease.

The elms of Lindridge were not spared. The local landscape never looked quite the same again.

~

I mourn enclosure, though the improving of cattle and sheep by selective breeding could not be done with the common flock, on common lands or in the strip-fields governed by the ancient, rigid three-course rotation.

Severely as it bore on the luckless, the Agricultural Revolution evolved a pattern of farming able, in some measure, to satisfy the now vast manufacturing population. The population of Britain in Tudor days was about three million; in 1800 it was over ten million.

The English Agricultural Revolution of the eighteenth century has become the story of great men, heroic individuals, whose names should be inscribed on the glad heart of every farmer: Jethro Tull, Lord Townshend,

Robert Bakewell, Coke of Holkham. These men are seen as having triumphed over a conservative mass of country bumpkins. They are thought to have single-handedly, in a few years, transformed English agriculture from a peasant subsistence economy to a thriving capitalist agricultural system, capable of feeding the teeming millions in the new industrial cities.

The role of these innovators is exaggerated. 'Turnip' Townshend, for example, was a boy when turnips were first grown on his estate. Jethro Tull was a crank, and his seed drill was not the first to be invented (a Venetian, Camillo Torello, took out a patent for a seed drill in 1566), and anyway was a mistake; his real objective was a horse-drawn hoe. Coke of Holkham was a great publicist (of his own achievements), but his 'Norfolk' four-course rotation of crops was already practised elsewhere in Europe.

Robert Bakewell's breeding experiments ended well with New Leicester sheep, badly with Longhorn cattle. The magnificent Longhorn was the quintessential triple-purpose English cattle, offering traction, milk and meat; Bakewell enervated them to the extent they were unable to survive standard farm conditions. The Collings brothers, hardly known outside the closed world of agricultural history, did selective breeding earlier and better, with the Shorthorn cattle breed.

That said, innovation and invention was in the eighteenth-century air. The processes were real enough, even if the dues of Tull et al. are pumped up. And they worked.

The seed drill of Jethro Tull allowed seed to be successfully sown in rows; Tull called his system Horse-Houghing Husbandry, the title of his book, published in 1731.

The Norfolk four-course rotation ascribed to Coke of Holkham was a handy guide for the general farmer. In the first year you grew turnips, which could be kept clear of weeds and well manured. These were then grazed off by sheep, whose dung enriched the soil. In the second year a crop of barley or oats was grown, on land that was now in a suitably high state of fertility; these were sown with clovers and rye grasses which remained after the cereals had been harvested. This pasture was grazed and cut for hay in the third year; in the fourth year the land was (theoretically) in exactly the right condition for autumn-sown wheat. The next year the course would start again.

The positive proof of the Agricultural Revolution is in the numbers. The expanding population was largely fed by home production. Wheat yields increased by about a quarter between 1700 and 1800 alone.

This new system of farming was remarkable because it was sustainable; the output of food was increased dramatically, without endangering the long-term viability of English agriculture. In a word, the system was organic.

Financially, the eighteenth century was the boom time for British farming. The landed interest succeeded in monopolizing the home market for wheat – the staff, and stuff of life then – via protection from imported grain, whilst securing bounties for the grain it exported. (The domestic cereal economy remained favourably biased until the repeal of the Corn Laws in 1846.) In this shielded economic atmosphere the British farmer thrived. He also had labour for what was, often quite literally, a starvation wage. It was not only enclosure that drove the farmworker to drink in Tenbury's pubs, or off to live in the Black Country.

~

Once, industry and agriculture were conjoined. Until the middle of the eighteenth century ploughs were made in the village by the blacksmith, but thereafter became factory products, made in cities such as Birmingham.

Craftsmanship and the maintenance of tradition were what mattered to rural society. Tools were ancient but beautifully preserved and there was a rhythm to life and a respect for Nature.

Loss in the countryside assumes many forms.

~

Quite aside from Coke, Bakewell, Tull and the innovations that bear their name, another reason for the productivity of English agriculture was my grandfather and his ilk: the farm manager.

So, this is the place to introduce my grandfather, more

fully, more roundly. Percival Amos, known as Joe to most and 'Poppop' to my cousins and me, was born in 1904. His parents were small farmers from Burley Gate in Herefordshire; the Amoses were proper yeomen (entering parish records as such in Herefordshire in 1617), surviving on small farms of forty acres despite the historical process of enclosure by Big Farmer.

Well, almost surviving. Small farms do not bear much division among descendants, so Joe Amos became first of all a farm worker, then a farm manager.

I toddled behind him as a boy so I saw his ability to command, the reverence inspired. He seemed to be everywhere; his workers called him 'The Whippet' for the speed of his walking, and this despite an injury sustained playing football for Burley Gate (the famous 'Robins' team), and suffering a hop-picking machine toppling on him. He limped noticeably.

A farm manager is the most conflicted of occupations: neither quite Us, nor quite Them. Before Poppop worked at Woodston he was a farm manager for the Prudential, who gave him a week off a year, one day of which was 'voluntarily' forfeited for a trip to the Hop Exchange in London, another day 'voluntarily' given up to show the London suits around the hopyards.

He did this effortlessly, being smart and smartly dressed: in a tweed jacket, cavalry twill trousers, Viyella shirt, green knitted tie, and the yellow waistcoat he wore when out fox hunting. Oh, and brown brogues on his feet. If it was raining, he wore his grey gabardine mac from Pritchards in Hereford, and his Hoggs tweed cap.

When I was four, I went on holiday to Switzerland with him. He wore the same ensemble to walk around Lake Lucerne, his mac neatly folded over his left arm.

~

The development of agrarian capitalism in England, with those involved in agriculture divided into landowners, tenant farmers, farm managers and labourers, saw the development of more systematic farm management and more efficiency in using the workforce. Spread of knowledge came through agricultural newspapers like *Farmer and Stockbreeder* (from 1889), and through shows like the Royal Agricultural Society's annual exhibition of stock and machinery. (The Tenbury Wells Agricultural Show continues to this day, and there is hardly a farm in Britain that does not take the *Farmer's Guardian* newspaper.)

Despite being 'trade', farming got a societal boost under George III, who liked his farming so much he was nicknamed 'Farmer George'. The Age of the Gentleman Farmer was ascendant.

Thus Woodston becomes more assuredly fixed in documents and on maps as 'Woodston Manor'.

~

What effect the Agricultural Revolution on one English farm, that of Woodston in west Worcestershire, going on east Herefordshire, going on south Shropshire?

Woodston followed suit; to practise the four-course rotation was to be agriculturally respectable; to question it was agricultural sacrilege. The Norfolk rotation held sway up to the Second World War at Woodston. I know, because my grandfather practised a variant of it.

Tull's drill was slow to catch on, but by the 1850s almost every farm of middling size or above was using an improved version of Tull's implement, along with the Shire horses to pull it. The horse increasingly replaced the oxen because the horse was one mile an hour faster, was nimbler around field corners than a lumbering ox-team of eight and did not require a tea break to chew the cud.

The Shire horse came to Woodston in the early 1800s – their size required the building of special stables – though the breed was then known simply as cart horse; the name 'Shire' was adopted in 1884. By then working oxen had all but disappeared from Britain. Yet, when a traction engine got bogged down, a team of oxen (if one could be found) was the preferred solution to the problem of rescue. Oxen were glacially slow, but they pulled with brute and steady strength.

Once, when I was ploughing with a horse, the blades turned up a rusted horseshoe. Except, when I examined it closer, I realized that it was the early Victorian footwear of an ox, a half-moon 'cue'.

Cattle were not specifically bred as work animals, but farmers and carters were always on the lookout for big, gentle bullocks which could be broken to harness rather than sent off to the butcher's blade. Oxen usually worked in pairs, and for life. Each ox had a name and within the pair one had a single-syllable name and one had a longer name. So Lark and Linnet. Gog and Gilgol. Traditionally, the left, near-side ox was granted the single-syllable praenomen, the right, off-side, the polysyllabic one.

The ox-shoeing process was somewhat rough, since oxen are disinclined to stand on three legs while the farrier fits the semi-circular cue to the fourth. First the beast was 'cast', thrown to the ground by ropes around its legs, then someone heavy sat on its neck to stop it getting up while other farmhands tied all four feet of the bellocking ox to a large wooden tripod the farrier carried with him.

As a beast of burden, the ox was undone by the railwayman of the 1830s. Suddenly it was possible for coal, iron, grain and other bulky materials to be transported at 30 mph by rail instead of 0.3 mph via oxen.

After two millennia of service, the ox was gone.

~

Just as the Penells dominated the history of Woodston once upon a Tudor and Stuart time, so the Adams family dominated the history of the farm during the Georgian and Victorian eras. By 1787, according to the Land Tax return, James Adams owned both Upper and Lower

Woodson. Born in 1751, James Adams's family were tenants at Stildon Manor, Menithwood near Lindridge, a property owned by the Catholic Blount family. Quite where the Adams money came from is uncertain – maybe marriage, maybe the coal industry, since Menithwood is on the edge of the Pensax coalfield. By the time James Adams died in 1806 he had built an archetypically Georgian house for gentry (redbrick, square, sash windows, on classical lines), thus replacing the Tudor homestead. He had also built a dynasty, and an empire: Woodston was approaching 3,500 acres in extent, the expansion realized by Adams buying out other farmers and expired leases. (In 1791 'Woodson' makes its first appearance on the map, featuring as a point of reference in the Kington to Stourport canal survey plan by Dadford. The canal was never completed.) Adams's heirs owned Woodston until after the First World War.

Adams was smart, on the button, thoroughly respectable. He was a churchwarden. (His brother, George, was perhaps the wild one; he was fined £5 at the Worcester Quarter Sessions, at Epiphany 1818, for using a greyhound for coursing. He died aged 39.) The Adamses had money, and the trappings to prove it: George kept a pack of hounds, and both brothers were devotees of horse racing. (The fifty years 1780 to 1830 were the golden age of the fox-hunting farmer, not least because they could hide hunt costs in the farm books and enclosure's hedges made perfect jumps; John Clare wrote witheringly about the farmers' new social mobility in the poem 'The Parish', where farmers' 'clownish taste aspires / To hate their farms and ape the country squires'.)

When the agricultural commentator William Pitt visited west Worcestershire in 1805 he was struck by the fact that 'the number of gentlemen who occupy land has increased considerably of late years, and some there are who hold forth very laudable examples of improvement'. He had men like the Adamses in mind and in view.

~

Wheat boomed during the Napoleonic Wars (1803 to 1815), so the new brick threshing barn erected at Woodston in 1815 is hardly a coincidence. The Adams family turned more of Woodston over to crops, notably the table-flat land between the Tenbury Road and the Teme, and switched from mainly sheep to mainly cattle; if the Adamses were not fattening cattle they were missing out on wealth, and since they increased their wealth, they were, *ipso facto*, fattening cattle.

The search was on for a big carcase for the urban masses. The local answer was the selective breeding of early maturing Hereford cattle, which replaced the Longhorn down on Worcestershire farms.

Breeding cattle for beef was a wholly English invention, one in which the meadows of Herefordshire and adjacent Worcestershire played a leading role. Until the eighteenth century, the cattle of southern England, having passed the medieval white phase, had become wholly red with a white switch, similar to the modern Red Poll. Breeders working on these red cattle of western England improved them into a distinct beef breed with a characteristic white face, the Hereford. Passing through Worcestershire in 1826, William Cobbett noted, 'The sheep are chiefly of the Leicester breed, and the cattle of the Hereford, white face and dark red body, certainly the finest and most beautiful of all horn-cattle.'

Nothing – not St George, rugby, cucumber sandwiches, cricket on the green – is as English as beef. It has been a national symbol for centuries, and for the French we are *les rosbifs*. The song 'The Roast Beef of Old England', penned in 1731, was once a national anthem, sung by the audience in theatres.

> *When mighty Roast Beef was the Englishman's food,*
> *It ennobled our brains and enriched our blood.*
> *Our soldiers were brave and our courtiers were good*
> *Oh! the Roast Beef of old England, and old English*
> *Roast Beef!*

There was enormous demand for Hereford beef, and drovers took herds to Birmingham, even London. Later, when the railway came to Newnham, just down the road from Woodston, the cattle went off to the slaughter yards of the cities in trucks.

The Hereford and the Shorthorn were Woodston's cows, down to the time of my grandfather; with my auntie Margaret, a girl of ten or so, he collected the cattle from the Round Market in Tenbury (auctions three times a week), and walked them along the A456 and A443. Six miles. In the 1940s you could do such things. There was barely a car or a lorry on the road.

Life in the countryside is small circles. Aged seven, I bought my first chickens from the Round Market, which was built in 1858 by James Cranston, and is actually oval, and has more than a hint of Chinese Gothic about it, with its pagoda-like roof.[2] An apposite place, anyway, to buy junglefowl.

~

Not all the sheep went from Woodston. The new improved Southdown breed was favoured by gentlemen farming their own estates, and the Shropshire Down – a cross between a Southdown and a local Midland or Welsh border sheep – especially favoured for its mutton, which was reckoned to be of far finer quality than that from New Leicester.

Britain was turning into a nation of meat-eaters, setting a trend to become common across the globe following industrialization. (Meat demand rises in line with higher incomes and urbanization.) Bakewell's New Leicesters possessed flesh aplenty; Bakewell took a long-framed, slow-maturing sheep and made it into a short-legged, hornless, barrel-bodied meaty one. He reduced the inedible parts of the carcase – the bone in particular – and in so doing produced more meat from each animal in the quickest time with the least food. Bakewell memorably declared 'all is useless that is not beef', meaning that the farmer's goal should be a sheep carcase with valuable cuts of meat, not wool or fat.

There were justified complaints that Bakewell's innovation had a heavy front end and ran to fat too easily. Bakewell responded with the pepper characteristic of the man: 'Mutton *anywhere* was welcome to the poorest classes.' When wool was king, the flesh of the sheep had been cheap meat. The animal was dispatched only after a useful working life as textile-provider and walking muck-spreader. So undervalued was mutton that fat from sheep for tallow candles was worth more than the actual carcase.

It was not until the eighteenth century that sheep were kept primarily for meat. Improving sheep for meat, however, had a disastrous effect on the fleece. England's great wool sheep, the Ryeland, became a pale ghost of its lustrous medieval self.

I know. I shear them. As often as not the fleece is only good for 'shoddy'.

~

The great stock-breeding revolution coincided with the introduction of oilcake, a byproduct from the extraction of linseed and rape-seed oil, first considered as a manure in the 1760s before farmers began to appreciate its value as cattle food, and soon experiments were going on to discover the possibilities of other 'concentrates'. Then the millers took over. Joseph Thorley of King's Cross, London, was one of the first in the field; he was already producing his compound cattle food at Hull in 1856. Another pioneer was the Kingston Cattle Food Company established in 1856, also at Hull. Its early advertisement for cattle cake referred to the 'highly condensed compound of pure, nutritious and fattening meals combined with valuable tonic, aromatic, stomachic, and gently stimulating agents'. Woodston had sufficient capital to pay for, and sufficient wheat to justify, a mill of its own. So, in the late 1700s, on the stream that fed the fishponds was erected a small watermill.

Even for successful farmers such as the Adamses of Woodston Manor, it was not all fun down on the farm in the eighteenth century. Livestock diseases ran rife. Between 1711 and 1769, ten million cattle in Europe died of rinderpest, a highly contagious and usually fatal viral disease. Rinderpest reached Worcestershire in 1748.

The practice of veterinary medicine had become commonplace under the Romans. After all, the stocks-in-trade of the Romans were farming and warring, and both required animals. The Latin for 'animal doctor' was *veterinarius*, as patented by Columella, who wrote twelve volumes on agriculture and animal husbandry *circa* AD 42–68. (His treatment for parasitic scab in sheep was olive oil, which likely had some efficacy.) Prescriptions by Roman wise men like Columella and Cato would be used in farmyards until Victorian times. Although the countryside of Europe swarmed with self-proclaimed animal healers, from 'cow-leeches' (who tended sheep as well) to 'hoggelders', the standard of their practice was epitomized by a medieval cure for constipation in oxen: a lively trout was taken from a nearby stream and committed to the gullet of the patient, under the assurance 'that it would soon work its way through all impediments, and speedy relief be afforded'.

Neither the ox nor the fish survived. But no matter because, ultimately, survival of beasts, like humans, was a matter for God. If the cow-leech or God failed, spells, astrology and charms were invoked. The most fortunate beasts until the middle of the twentieth century were those left untreated, because their peers were purged, bled, fired – or stuffed with trout – before dying. As a boy, my grandfather witnessed cows and ponies being bled to 'let the badness out'.

In 1940, while supervising the resiting of a wooden loo for the seasonal hop-pickers at Woodston, Poppop uncovered a mass pit of charred bovine bones. These were the remains of cattle slaughtered because they had contracted rinderpest or some other lethal bovine disease.

~

I found the lump yesterday morning. She was leaning forward, taking a drink of water, and my hand was on her neck, caressing her.

And there was the lump. It was dark, so I felt further down. Another lump.

It is what every farmer fears, a cow reacting to the skin test for bovine TB.

Oh, the principle of the TB test is simple enough: each cow is injected in the neck by the vet in two places, first with an antigen from bovine TB, secondly with an antigen from avian TB. Seventy-two hours later, said vet comes out to check for bumps. A 'reactor' cow, one carrying the disease, will come up in bumps. Which is a death notice, delivered by a green tag punched in the ear.

Except . . . the assessment of bumps is far from simple. It depends on how precisely big is the bump. Then, is the avian or the bovine bump bigger? Some delicate, wilting-violet cows will get bumps from the steel needle, let alone its active contents.

The merest hint of a bump on a cow and I panic. Reason flies out of the window. I fluster. I cannot remember even whether the key skin prick, the bovine, is top or bottom.

Every cow has worth, but in a small herd of ten like ours, every cow has meaning too. Plus, I confess, a name. In the case of our Red Polls, these by family law begin with 'M'. (The book of baby names has been well plundered over the years. We've done the spectrum from the haughty Margot – think a cowy version of Penelope Keith in *The Good Life* – to the pneumatic Mirabelle, a natural to advertise butter.)

Milly is a particular favourite, with her fluttery eyelashes and pawky humour. A couple of years back she kept escaping, and I was utterly perplexed until I caught her red-hooved. Her modus operandi was to squiggle on her side under the single-strand electric fence.

I wouldn't want to lose her. Really, really wouldn't.

I tell you now, the TB test is hell. I don't sleep before it, don't sleep during it.

Today is D-Day: the day the vet comes back to check for 'bumps'. I got the cattle penned for examination just after first light, and getting the girls into the pen is sufficiently stressful for me to pop 75mg aspirin pills. (I read somewhere once that aspirin mitigates strokes. Or heart attacks. One or the other.) It is stupid, but one does not want to be the cowman whose failure to pen cows is tittle-tattled by the vet down the line to the neighbours.

I took no chances. I led the girls into the pen with a shaken bucket of cattle cake, me running faster than anyone at Pamplona, because being caught up by 5 tonnes of milling beeves is interesting to say the least. The girls in the pen, I two-arm-vaulted the side, nipped back around and closed the gate on them.

Works every time, except the time I failed to clear the pen's side, fabricated from motorway crash barriers, and broke three ribs.

A really great cowman, like our old neighbour Jack Williams over at Abbeydore, can pen cattle, even Limousins, by hypnotizing magic. He sang them in.

It is almost ten o'clock now. The vet is due, and I'm by the pen, paralysed with worry, mouth sticky-parched as when you take a Communion wafer.

The mist is low, axeing off the top of the trees beside the pen. There is no compassion in a tree; on a day like this their unthinking insensitivity, their self-regarding composure merely irritates.

Somewhere off in the hedge, an unseen robin drips sad soliloquies.

Lower still comes the mist to lie on the backs of the staring cows in white shrouds.

The tweed-jacket era of James Herriot is long since gone. Bob Minors looms towards me in green waterproofs, a doppelgänger for an alien in *Dr Who*, *circa* 1980.

'Any problems?'

I grimace. Then push the internal gate behind the cows, so they are pushed into the narrow metal-railed corridor of the 'race'. (An odd noun for the place of the slow queuing of animals.) To touch, the gate has condensed the chill of the centuries in its bars.

Cows are hierarchical. Milly is mid-herd, number 6.

Bob leans over, runs his hands over the injection sites on the cows' necks. He reaches Milly. 'Ah,' he says. And gets, from out of his toolbox, the calipers.

He squeezes her warm vital skin and pinches it with the steel calipers. Once, twice, thrice, he measures the bumps on Milly's neck.

Somewhere across the valley comes the coffin-closing croak of a raven.

'Close, but good enough,' says Bob.

I do not understand, comprehend, and ask him to repeat.

'She passes.'

The world blurs. He runs his hands over the remainder.

'They're yours for another year. I'll see myself off. You let them out.'

I release the cows from their confinement, and me from my anxiety.

We run around, the cows and I in the mist, laughing.

~

By the roll of the nineteenth century, the man and woman in the city were becoming as fond of a pint as they were of a bit of beef and a slice of bread. Lindridge started to become the epicentre of Worcestershire hop-growing. Passing through Lindridge in 1805, William Pitt, the agriculturalist rather than the prime minister, noted 'hop yards numerous in sight of the road'.[3]

With their moistened fingers to the wind of change, the Adamses invested heavily in hops, turning Woodston into one of the main hop farms of England.

Daphne (left) and Kathleen Amos sorting hops at a 'crib', Woodston, c. 1950.

CHAPTER VII

THE BINES SEEK
THE SKY

*From the depression at the end of the Napoleonic Wars to
the Edwardian era – The century of wheat bust and
Woodston's hop boom – Mechanization and the 'Swing
Riots' – High Farming*

*In valleys of springs of rivers
By Ony and Teme and Clun
The country for easy livers
The quietest under the sun*

'A Shropshire Lad', A. E. Housman

THE TRADITIONAL TOOLS FOR harvesting are the scythe
and sickle: the scythe predominant, being all-purpose,
efficient on long grass as well as corn; the true sickle,
developed for corn, has a serrated edge.

Scything, or 'sickling', is back-breaking manual work.
I know because I scythe a five-acre field.

The first machines to mow a meadow and reap a wheat field were comically inefficient. Credit for the first practical mechanical reaper is usually granted to Cyrus McCormick, a Virginian farmer who produced his machine about 1840. The design was greatly improved by further patents in 1845. By 1879, McCormick had won a gold medal from the Royal Agricultural Society. The patented McCormick reaper could be used for grass as well as for corn.

You'd expect the farm labourer to have been cheered by this mechanical salvation from horny-handed drudgery. He was not. There was artisanal pride in good reaping, the effortless flow back and forth of the swishing blade, the making of hay by hand. In scything, too, a zen-like state of contemplation was reached, a sort of satisfied philosophy. These are my notes on making hay the old way:

Haymaking follows the Goldilocks Principle. You need a day that is 'just right', not wet, not scorching either.

On the green lane to hay meadow I see a green woodpecker, the rain bird. (Her flight is up and down, as if she surfs invisible waves.) In my mind, I hear my French-speaking stepmother say: '*Lorsque le pivert crie / Il annonce la pluie.*' Sure enough, despite the Met Office forecast, a shower comes from out of the eggshell blue sky. No haying today.

Two days later. 1.15 p.m. Back at meadow. Muggy. Almost all bird song has stopped; the July drought of avian sound. Moulting birds/nesting birds do not wish to advertise presence.

Mark, with two prongy hazel sticks billhooked from hedge, 'plot' where a meadow pipit is nesting. The nondescript meadow pipit is a true bird of meadowland.

Walk the edge, survey the grasses, yard high. Most gone to seed, and beyond. Empty purses on stalks. Would have been better as fodder if it were protein headed, but cutting when wildflowers have dropped seed means those wildflowers have a chance to reproduce. Let's give Nature a chance.

Stumble upon a patch of bird's-foot-trefoil, gold interwoven with red and so dense it is a doppelgänger for embroidered kneeler in church.

About to start mowing with the tractor: common blue butterfly flickers around white plant I cannot identify. Open tractor toolbox to get wildflower guide, stop. The butterfly doesn't know the plant's name but enjoys it (happy in heedlessness), so why do I need to nomenclature the plant, pin it with scientific precision?

Butterfly splashes around in a puddle of sunlight; *cf* child in a paddling pool.

Some time after 3 p.m. Been up & down, up & down on tractor, clickety-click cutting with the bar mower. Drink break. Three-plus acres done. The cut grass is in gorgeous green braids along the meadow, with yellow from buttercups, red from clover and white from yarrow tied in.

Demeter always wears flowers in her hair. Something pagan, vital about olde time hay. No wonder peasants wanted 'a roll' in the stuff. Not much of a euphemism, is it?

Vanilla attar of vernal grass, the Turkish Delight smell of red clover.

Circa 3.45 p.m. Pale, and shaking. Bottom of field steep, slightly slippery. Almost overturned tractor. For a moment was on the two right-hand wheels. Suspended. Hanging. Frozen. Then a desperate flinging of weight to counterbalance. Sick.

Words like 'Discretion' and 'Valour' whizzing to mind. 'Cowardice' too, possibly. Decide to cut bottom last half-acre with the Austrian hand scythe brought for field edge under the trees.

The knack with scything: keep the blade flat to the ground, a mere millimetre off the surface. Swing scythe around body in a circular arc. Forget grunting Poldark. A man scything should be mistaken for a quiet man performing Tai Chi.

Robert Frost in 'Mowing' declared scything to be 'the sweetest dream that labour knows'.

My long scythe lullabies to the earth. Hush. Hush.

Then: *shwhit-shwhit* of the whetstone on the blade, as I hone it. Rasping note of steel to echo the chiff-chaff in the hedge.

I try to remember medieval calendar verse for July: 'With my scythe my meads I mow'?

The repetitive swish of the scythe as it razes the towers of feather-headed/ bobble-headed grass is producing of pontification.

At a certain sideways angle, the meadow top is a red mist of sorrel spires.

Strange small holes revealed in the earth. Shrews?

Reptilian slither: a grass snake S-coils, spurts away to the hedge, causing an incongruous sss-shiver on this sss-ultry day. Sss-nakes in the grass-ss, indeed.

Chew on a leaf of sorrel. Acetic, saliva-stimulating, thirst-slaking; sorrel was always the Haymaker's Friend.

Out of the eirenic haze, horseflies (slate flat & slate grey) come in battalions down to this tree-sheltered, hedge-snug, low end of meadow. In the field next door, the Lim cattle become agitated. The flies attack them. Dozen cows gallop across the pasture, tails up.

The scythe exposes networks of vole tunnels in the grass. Tramp stink of vole pee detectable to fox noses; also ultra violet & detectable by kestrel eyes. Double jeopardy.

Decapitate top of vole nest. Inside two naked vole babies. Sugar voles. (Re-cover them with red earth.)

Hotter, as the afternoon wears on. I wear down.

Some inkling makes me lift the blade as it goes into a tussock. Inside, a bundle of white fungal balls. But not mushrooms. The asp was the clue: they are eggs of a grass snake. A thing never before seen by me.

If you want to know about the natural life of a meadow, ask the fellow who got down off the tractor and cut the grass by hand, so that the meadow was close-up and personal.

~

The reaping machine, with its giant scissors on the back of a little horse-drawn buggy, devastated wildlife on the farm. Whereas the hand-scyther would cut around the home of the leveret, the nest of the corncrake, the nest of the grass snake, the unfeeling machine rode over all with spinning steel blades, and chopped harelet and embryonic corncrake and snake egg into red and yellow bits. The farmer sat up on the reaper, away from the ground, thus beginning the disconnect that has concluded with the farmer inside an air-conditioned tractor cab, the tractor's progress through the field determined by computer.

English farming, which produced the landscape for so many species to thrive, now killed them off. The death-drop of the skylark began in 1840. I once talked with Poppop about skylarks. He said that over a hay meadow scythed because it was too steep for a horse-drawn reaper, this being about 1920, he saw more than thirty skylarks in the air at once, their song making a continuous canopy of trilling.

A portrait of the wildlife paradise lost on the English farm is left us by Wiltshire farmer's son Richard Jefferies, who wrote in 1834:

All the village has been to the wheat-field with reaping-hooks, and wagons and horses, the whole strength of man has been employed upon it; little brown hands and large brown hands, blue eyes and dark eyes have been there searching about; all the intelligence of human beings has been brought to bear, and yet the stubble is not empty. Down there come again the ever-increasing clouds of sparrows; as a cloud rises here another cloud descends beyond it, a very mist and vapour as it were of wings. It makes one wonder to think where all the nests could have been; there could hardly have been enough caves and barns for all these to have been bred in. Every one of the multitude has a keen pair of eyes and a hungry beak, and every single individual finds something to eat in the stubble. Something that was not provided for them, crumbs that have escaped from this broad table, and there they are every day for weeks together, still finding food. If you will consider the incredible number of little mouths, and the busy rate at which they ply them hour by hour, you may imagine what an immense number of grains of wheat must have escaped man's hand, for you must remember that every time they peck they take a whole grain. Down, too, come the grey-blue wood pigeons and the wild turtle-doves. The singing linnets come in parties, the happy greenfinches, the streaked yellow-hammers, as if any one had delicately painted them in separate streaks, and not with a wash of colour, the brown buntings, chaffinches — out they come from the hazel copses, where the nuts are dropping, and the hedge berries turning red, and every one finds something to his liking. There are the seeds of the charlock and the thistle, and a hundred other little seeds, insects, and minute atomlike foods it needs a bird's eye to know. They are never still, they sweep up into the hedges and line the boughs, calling and talking, and away again to another rood of stubble without any order or plan of search, just sowing themselves about like wind-blown seeds. Up and down the day through with a zest never failing. It is beautiful to listen to them and watch them, if any one will stay under an oak by the nut-tree boughs, here the dragon-flies shoot to and fro in the shade as if the direct rays of the sun would burn their delicate wings; they hunt chiefly in the shade. The linnets will suddenly sweep up into the boughs and converse sweetly over your head.

I read this extract at least once a year, to remind myself that the way to the future is the path to the past. It is called organic farming. We can have our cake and eat it, birds, beasts, and humans all. As a proof I submit my diary extract, from the early twenty-first century. But Richard Jefferies would be comforted by its familiarity:

I went down to the field beside the brook because tomorrow was the end.

The corn-glow of late evening was warm on my neck, and the dog at-heel panted, tongue-lolling. In the sky, the great icebergs of the cloud were becalmed.

We took the short cut through the wood where the leaves, thickened by the lignin of ageing, provided shade as efficient as a gentleman's black brolly; in the oaks, at least, it was cool.

We scrambled down the path, out into the open, and there below was my personal golden ocean, ten acres of it, unmoving in the heat, heavy with its own ripeness, but also heavy with the tiredness that seems to exist within everything in August. When summer's lease is over.

Running the whole breadth of wheat field, from edge of trees to edge of water, were just discernible lines, like the watermarks on paper held to the light: roe deer tracks.

At the far end of the field, Vs of deer heads poked above the yellow surface of the sea; the bodies of the beasts were drowned from view.

There was a certain anxiety about the visit, as there always is before a crop is safely gathered in. Anything could have happened since the morning check-up. A plague of locusts. Freak wind. Crop circles. In my life I have seen yards of hops lost to mildew, fields of potatoes perished by blight, barley battered to doom by hailstones.

But all was good yesterday evening. On reaching the shore line, I reached in and plucked a head of wheat, pulled it apart, split a single grain between thumbnail and forefinger. It was hard like a nut, like a gritstone. Which is how ready-to-cut wheat should be. Over the field too was a sense of expectancy, such as you get with a crowd waiting for a football match to start.

Tomorrow the Claas combine would come, weather willing. And the portents of meteorology were good. Behind me, in the sky above the disused village quarry, a swatch of red, which is the promise of a good dawn for the harvester.

In the sticky air swallows lacily hunted insects. Their hirondelle compatriots, the swifts who nest on the church tower, left last week. The coming of the swallow marks the beginning of summer; the departure of the swift summer's finish. The birds mark the turn of seasons more meaningfully than the calendar.

I wondered for a moment: Were these the same swallows as last year? Always I feel one should be able to tell the individual birds apart, and it was a failure to think 'swallows' as though they were a uniform collective.

Close up, the field was not pure gold, and that was good, because the crop was organic, and thus mosaiced by arable wildflowers, or what some would term 'weeds'. Scarlet pimpernel, cornflower, scabious, groundsel, white campion, shepherd's purse . . . and twenty others. A few poppies lingered on from spring, blaring their existence by their blood-vividness. A poppy in a cornfield always recalls to mind JMW Turner's 1840 seascape with red buoy; the irresistible attraction of scarlet.

But tomorrow the combine would come, and they would be cut down. All the wildflowers.

Close up, the field was hissing with sound, not the writer's sibilant 'rustle of wheat in breeze', for there was no breeze – the air was as solidly calm as glass – but with the stridulation of grasshoppers and the buzz of bees. Across the little stony brook, through the alder trees, a neighbour's conventional wheatfield . . . The summer was silent there.

I walked on, along the grass ride by the alders of the brook, and the deer, finally awakened from their devouring, saw me. For a moment, they stood stock still, deliberating, then privileged safety over stomach and ran for the woods through the corn, with that particular undulating rhythmic bound they have. Wave riders.

On an impulse I did not at first understand, I decided to lie down on my front and peer between the stalks of the wheat; Gulliver in Lilliput. The dog settled next to me and almost immediately went to sleep, with the easy somnolence that Black Labradors possess.

We were there for a while, and gradually realization washed over me. In lying down, I lost the haughty, human perspective. Got down with Nature. Felt the terror of small things.

There seemed to be no end of creatures, creeping, crawling through the dense alleyways between the wheaten spires; and sometimes up and down those same golden cellulose towers.

A field mouse clambered up a stalk to the mitre head of grain; the stalk bowed, broke; back down on earth, the mouse pulled apart the grains with mouth and its pink hands; which are those of tiny people. A toad, warty and portly, waddled through the scene. Some goldfinches chinked prettily.

The dog snoozed on. The landscape lay in its swoon.

At one point, I became aware of a ripple in the atmosphere. I rolled to my side, and looked up. A barn owl was passing overhead on silent wings, its breast feathers the same blinding white as the clouds. The owl turned its head in a slow emotionless wheel, as if it were a flying robot, regarded me, noted me, flew on.

I was too big to eat, and of too little interest for anything more than a passing glance.

In my humility, I carried on watching the life of an arable field. There were beetles, butterflies, moths and roving common harvestmen, those stilt-legged predatory arachnids that look like spiders but are not. One harvestman scuttled about on six legs, instead of its Evolution-ordained eight. *Phalangium opilio* have detachable limbs, so when they themselves get caught by the legs, they jettison the trapped limb to hunt another day.

In the end, I had to leave, though I had no wish to. I prodded the dog awake. As we stood up, I saw that the owl was still quartering, but very low now, skimming the surface of the ocean. And that its back was the same sand-gold as the corn.

Tomorrow, the combine would come, and I thought, All this will be gone. In my head, the elevated processes of ratiocination reassured me that the field had done its bit to preserve the creatures and the weeds of ploughland.

But a sadness, still.

～

Almost gone now, the practice of winter stubble, of leaving the cut-cornfield untouched till spring . . .

When my parents divorced, I spent spells of my childhood living with Poppop and Grandma in their house at

Withington. They would receive Christmas cards showing scenes of pheasants and partridges perambulating the snow-crusted ruins of the cereal crop. I would look out of the bathroom window on to the field behind and see the same scene. Since the great switch in the 1970s from spring to autumn ploughing on the arable lands of Britain, however, over-winter stubble is hard to find. Only 3 per cent of arable land is left as winter stubble. Aside from giving the land a rest, stubble was platter and refuge for the wild things. A lack of winter stubble, with its spilt grain and its weeds, is a cause of decline in our farmland birds and animals.

And, yet, so easy to remedy.

> *The little mouse comes out and nibbles*
> *The small weed in the ground of stubbles*
> *Where thou lark sat and slept from troubles*
> *Amid the storm*
> *The stubbles ic'el began to dribble*
> *In sunshine warm*

From 'Address to a Lark Singing in Winter', John Clare

~

First light. Below, down in the village, a cockerel crows. Across the millet stubble, in the black wood, a tawny owl yaps. Otherwise, a world of silence.

The stars are still alight, alchemizing the puddles – which sprawl around the geometric precision of the straw spikes – into silver mercury. Noughts and lines. There is bleak, binary beauty in a stubble field in midwinter.

It is breath-blowingly cold. First light is a strange time of day to be dog training, but when otherwise is the time? When? I have found there to be too much comedy in teaching (or, as we say in Herefordshire, 'learning') a black Labrador at night. In these very first monochrome moments of a winter's day, she is at least faintly discernible.

The millet stubble comprises twenty acres, but the field is thin. So, it is a long walk along the length of the rectangle to the wood. Perhaps five minutes. In my left gloved hand a small portion of cheese.

On the command 'Heel!', the dog and I set off across the stubble, this endless plot of pale points, towards the dawn. Our own bed of nails. The dog, Plum, glue-sniffs my dangling hand.

The stubble is eight inches high; under my wellingtoned feet, it sloshes and crunches, snow-like.

Directly ahead, the fuzzy felt wood; to the right a long low hill, like the cliff of a continent seen from sea in the dark. A car comes down the hill, a commuter, its headlights blinding across a mile or more. For a flashing second we are illuminated in no-man's-land. Stark statues.

Unfrozen by the car's turning direction, we go on.

We have been striding for two minutes when it happens. Out of the earth, birds begin to shoot up. Chaffinches I can distinguish in the gloom from shape, flash of white on wings. The other birds . . . I guess at tree sparrows.

Every press of my feet on the earth releases birds. And squirts of bird song. I hear linnets, I hear finches, buntings. The only useful analogy I can find is – walking on whistling sand.

I press extra hard, and there is a land mine of an explosion. Three red-legged partridges blast off, wing-whirring away behind us, back into the night.

Plum stays steady.

Half way. 'Good dog.'

On towards the advancing tide of daylight. Homer talked of the 'rosy-fingered dawn'. How heatedly Hellenic. In December in England it is cold, 'pale-digited dawn'. Still, there is not a cloud in the sky.

Then, I see a sight I have read about but never encountered: a brown rug rises from the ground and, as though taken by a gust of wind, flaps to land hesitantly twenty or so yards away.

I stir the soup of memory for the collective noun for skylarks. A gobbet surfaces: 'exaltation'.

There must be fifty or more skylarks in this winter flock.

By now, the dog and I three-quarters of the way across the stubble, the daylight sheening on the stalks, her back. In a long wet mirror on the floor I see: a man and dog. Nails riven through their image.

Three red deer stop their grazing, raise their llama necks – then bolt for the wood. The daylight touches two clods of earth into life, and they too run for the wood. Brown blurs. Hares.

The wood; at night it exhales its animals; at dawn it inhales them in.

The dog and I have now gone from the world of black-and-white into the world of colour. The stars . . . they have melted away. Under my feet, the emerald green of arable weeds: groundsel, mouse-ear, docks, thistles. The bright yellow of hawkbit.

We complete our traverse, reach the stone track. I open my hand, the dog takes the cheese. A dog's nose in the hand; it is a familiar feeling, perhaps twenty thousand years old.

'Good dog,' I say, 'good dog.'

This is what happened on a six-minute walk across organic millet stubble in December, the month when the year ends, and is born again.

~

The new-fangled machines of the Agrarian Revolution put people out of work, as well as mangling skylark nests. With the end of the Napoleonic Wars in 1815 (Waterloo and all that) came a collapse in the prices of farm goods, and, inevitably, because woes always go hand in hand with other woes, there were bad harvests. Old Cobbett, on his journalistic rural ride to divine the state of the nation, reached Herefordshire in 1826, to find:

> As to crops, in Herefordshire and Gloucestershire, they have been very bad. Even the wheat here has been only a two-third part crop. The barley and oats really next to nothing. Fed off by cattle and sheep in many places, partly for want of grass and partly from their worthlessness. The cattle have been nearly starved in many places; and we hear the same from Worcestershire.

Farmers, finding their profits dissolving, attempted to correct the situation, partly by reducing wages, partly through Parliamentary legislation (notably the passing of the Corn Laws of 1815, which sought to maintain farming prosperity by restricting imports), partly by introducing labour-saving machinery, such as the threshing machine. It was this device, invented by Andrew Meikle of East Lothian in 1786, which occasioned the labourers' despairing protest of the autumn of 1830, sometimes called the Swing Riots.

The average rioter was young, male, unemployed, receiving little relief and often forced into paltry-paid parish work like mending roads. If in work, his wages had fallen (to about 2s. a day from 3s.). Fewer labourers were employed all year round. On 6 December 1830 the Swing Riots skirted Woodston, when five young men at Hanley William, a mile across the Teme, were arrested for destroying a threshing machine left at the roadside, the first to be used in their parish.

This is Woodston in west Worcestershire, peaceable and well-ordered, even well-off. The rest of the country was in a mayhem of machine-breaking.

The Hanley boys saw the threshing machine at Hanley William as the precursor of many. Rightly so. Within a generation, steam power would arrive on the Worcestershire farm, and by the end of the century traction engines and threshing machines were a countryside commonplace, usually owned and run by a contractor, a spivvy middleman, self-important in the reflection of his own polished metal. Another alienation of the farmer from his own field.

The magistrates, with the wisdom of the years and knowledge of local life, dismissed the charges against the Hanley Five.

I wonder about the Hanley boys. Ne'er do wells or brave principled lads? My grandfather, Joe Amos – conservative, middle-class, a monarchist who once asked my mother in her sixties to leave the house because of some mildly disparaging remark about the royal family – led a strike at his school when he was thirteen.

～

Rage against the nineteenth-century machine achieved nothing, and the flight from the land for the supposedly bright gaslights of the city persisted. The numbers of workers involved in agriculture dropped, dropped, dropped some more. By 1850 only 22 per cent of the British workforce was in agriculture; the smallest proportion for any country in the world.

The long Victorian century, in agriculture, went from bad to worse. A run of unfavourable harvests, from 1876 to 1879, and cheap imports from Canada, saw the price of wheat drop from 58s 8d a quarter in 1873 to 26s 2d a quarter in 1896 (the nadir of the depression). The general price index for farm products was almost halved.

Then there were the plagues of the animals, which visited with medieval abandon.

~

In 1839, an article appeared in *The Veterinarian* by Mr Hill, veterinary surgeon of Islington Green, warning of a disease attacking cows in London dairies. The clinical signs were 'inflammation and vesication' (blistering) of the mouth and 'a continual catching up and shaking of one or other of the hind-legs'. People initially called the disease 'the vesicular epizootic' or 'malignant epidemic murrain', before settling on 'foot-and-mouth disease'. It ripped through the country, and a despairing Royal Agricultural Society of England agreed to subsidize a Pathological Chair of Cattle at £200 per annum at the London school.

The job advertisement required a person 'of some education – of more talent – of long experience in cattle practice, not past middle of life, and all his faculties unimpaired'. Twenty-nine-year-old James Beart Simonds, the first incumbent of the chair in Cattle Pathology, had all the attributes. He carried out a crucial transmission experiment. He wiped a handful of hay over the face of an infected animal, then fed the hay to a healthy cow – in which the disease appeared: *ipso facto*, the disease could be spread by contact between beasts. Simonds' experiment also showed that diseases in farmyard animals could be controlled by culling or movement restrictions, even if they could not be cured.

The calls on Simonds' expertise came thick and fast, but two cases, twenty years apart, were landmarks in veterinary history. In 1847, a sheep farmer in Datchet invited Simonds to view 'a peculiar eruptive disease' in his sheep. Simonds, on visiting the farm, diagnosed sheep pox, and traced the outbreak to merino sheep imported from Saxony. At Simonds' behest, the government eventually passed an Act of Parliament in 1848 'to prevent the Introduction [to Britain] of contagious or infectious Disorders among Sheep, Cattle, horses and other animals'. And in 1865 a Mrs Nichols called Simonds to her dairy in Islington, London, where large numbers of cattle were dying. Simonds quickly diagnosed cattle plague, the first outbreak in Britain for nearly a century. From personal experience on the Continent, Simonds knew that treatment was useless, and again recommended, in the absence of a known cure, a 'stamping-out' policy, the restriction of movement and the cull of infected beasts.[1]

His drastic recommendations found no favour with

the cattle trade or a government given to laissez-faire. The Archbishop of Canterbury prayed for help from the Almighty: 'Stay, we pray Thee, this plague . . . shield our homes from its ravages. Amen.'

The congregation at St Lawrence's, on the hill above Woodston, also doubtless went down on their knees for divine assistance.

Only after 400,000 hapless cattle had died did the country heed Simonds' advice, and in February 1866 the Cattle Plague Prevention Act was rushed through Parliament. Cases went from 18,000 a week to eight a week in nine months.

With cattle pox rampant, and wheat 'not worth a light', how did Woodston boom?

By following the hop.

～

The scientific name for the hop, *Humulus lupulus*, comes from *hummus,* earth, and *lupus*, wolf, the latter referring to the plant's habit of attacking and strangling its host. The English name is more prosaic, and comes from the Anglo-Saxon *hoppan*, meaning to climb.

Today, the names 'beer' and 'ale' refer to much the same drink, but the word 'ale' was originally reserved for brews produced from malt without hops. This was the original drink of the Anglo-Saxons and English, whereas 'beer', a brew using hops, originated in Germany.

The plant is native, but was not cultivated here until

the end of the fifteenth century. Our national drink until then had been ale, unhopped and frequently flavoured with herbs. Brewers started to import dried Flemish hops but these contained so much extraneous matter that an Act of Parliament was passed in 1603 imposing penalties on merchants and brewers found dealing in hops adulterated with 'leaves, stalks, powder, sand, straw and with loggetts of wood dross'. The point and purpose of hops was to 'bitter' and to preserve. By the mid-seventeenth century 'a pint of bitter' beer was the established drink (not just as 'porter' for the proletariat, but as 'pale' for the bourgeoisie), and in a successful year, an acre of good hops could be more profitable than 50 acres of arable land. The breweries of Burton-upon-Trent wanted hops, and Woodston had hopyards of them: There was Great Hopyard, Vicarage Hopyard, Lower Field Hopyard, Monk Hopyard, Skimmetry Hopyard and Barley Close Hopyard.

Woodston was one of the first farms in Worcestershire to grow hops, and by general acclaim it turned out to have some of the best land in England for the crop; the red Devonian marl, fabricated in Earth's bowels a billion years before, incorporated the right minerals, the right pH balance and the right structure to hold the plant's roots as it climbed for the sky. Woodston had the right climate too: enough water, but not too much; enough sun, but not too much. In the modest mile-width of the Teme valley, the low comforting hills protected

Woodston – keeping in warmth and keeping out wind – like encircling arms.

Hops were planted at Woodston by 1700. They became a farmer's favourite, an English people's too. To prevent adulteration of their preferred pale beer, an Act was passed requiring the bags or 'pockets' in which the hops were packed to be stencilled with the year, place of growth and the grower's name; a tradition that continues to this day.

By 1840, there were some 6,000 acres of hops in Worcestershire; in 1894, the peak year, over 10,000 acres, and about the same amount in neighbouring Herefordshire.

It was the golden age of the hop industry; at Woodston new kilns, for the drying of the hops, were built in 1880 in red brick, the size of a castle, their four white cowls visible for miles away as they turned in the wind. Some said they were the biggest kilns in the whole of Albion. Providentially, there was a ready supply of coal in the Mamble and Pensax pits at the edge of the parish, small drift mines, worked by a couple of men, but fuel enough to turn into coke for the kilns. The farm now had some 150 acres down to hops. To protect the hopyards, the old enclosure hedges were allowed to grow to 20, 30 feet high. The birds loved that. (I know because my playground as a child was a hopyard, and every one of the four green walls rustled with the activities of passerines, as if a breeze was running through them, always.)

Another proof of Providence was that the frequent

flooding of the Teme actually benefitted the hopyards, depositing a layer of silt, which was nourishment.

~

Life in the countryside is indeed composed of small circles. Spirals, even. One of the pit-owning families at Pensax were the Yarnolds, with Samson Yarnold listed, in 1871, as the last owner-worker of a mine in the coalfield. My aunt Margaret of Woodston Farm married Tom Yarnold in 1955.

~

William Pitt (1749–1823, and no relation to the political Pitt dynasty) was one of those employed by the Board of Agriculture to prepare reports on the state of agriculture in each county. Among Pitt's reports was *A General View of the Agriculture of the County of Worcester: with observations on the means of its improvement*, published in 1813 but written 1805–7. In its pages Pitt provides a full picture of hopgrowing at Lindridge, noting its need to 'be well manured with good rotten dung, or compost', and that, typically, beside the Teme, planting was one thousand 'stocks' (plants) per 4,000 square yards. About 5 hundredweight of hops was given up by each acre. The hopyards also 'doubled up' with other crops, particularly fruit trees, potatoes and turnips. (Compare and contrast with late-twentieth-century agriculture's mono-cropping.) The main hops grown were Golding-Vine and Mathon-White.

Pitt continued:

The first year green crops . . . are cultivated on the
skirting of the ridges, and the hop-plants are three
times kerfed [a kerf is a hoe with a steel blade as
much as 9 inches wide], to mould and keep them
clean, but are not poled; the second year they are
poled about May-day, and may produce half a crop;
the third year, and afterwards, they are supposed in
perfection.

Poles, Poling, Tying. Two poles are used to each
stock, two thousand to an acre; they cost 8s. per hundred, and about two hundred upon an acre annually,
or one-tenth of the whole, are reckoned to wear out;
they are tied to the poles with rushes, at 4s. per acre.

The Distempers, to which hops are subject, are
blight and mildew; for the prevention, or cure, of
which, no remedy is known; they are supposed
entirely dependent on the seasons, as the crop is very
precarious, and subject to changes so sudden, as to
baffle all human care, or foresight. In the month of
July, a sudden blight has been often known to raise
the price of the stock in hand in a few days, sometimes to double their former value; and the dealer,
who has before been pushing off the article, in
expectation of the price lowering, has, on a sudden,
locked up his warehouse, and refused to supply his
customers.

Picking. This is done by women and children from the neighbouring populous counties, or from Wales, principally in September. Women hop-pullers have 8d. per day and breakfast, or 9d. without, and three pints of beer, or cyder, each per day; eight pluckers to a crib in three days pull an acre; expense, 18s. to 24s; but this depends upon the bulk of the crop in some measure . . .

The hops were picked into 'cribs' (wooden frames 9 feet long, 4 feet wide, 4 feet high supporting a cloth bag), then, when full, were transferred to cloth sacks to be taken on a horse-drawn cart to the kilns, to be dried out over coke fires.

The Expense attending the hop culture is pretty considerable . . . this is to be repaid by the green crops attending it, or by future profit.

Ordinary Expenses, such as cultivation with plough and kerf, tying, poles (2000 to an acre), drying, duty (at 2d, per lb. suppose 6 cwt. per acre), manure, work out at £15 per acre.

Pitt estimated return per acre to the grower to be £61. Adding:

That there is a fair profit to the grower, may be supposed, otherwise the culture would cease; but, it is very certain, that much more has been gained by speculators in this article, than by the growers; what has been gained by the latter, has been chiefly by men of property, who could bear stock, and keep their produce till it would bring a good.

Most of the estates which grow many hops, have plantations in which the poles are raised. Ash and barked oak are preferred; but willow, poplar, and alder, are also used. Where the estate does not produce a sufficiency, they are bought at the woods and coppices in the neighbourhood, at from 5s. to 15s. per hundred: their length is from eight to eighteen or 10 to twenty feet, proportioned to the goodness of the lands; they last, with care, seven or eight years.

The tythe of hops is more particularly complained of than that of any other article, and considering the very great expense at which they are cultivated, it appears to be with reason. The present regulations respecting the hop duty are not complained of; and if the tax must be continued (to use the language of the planter) it cannot probably be altered for the better; the only use of consequence to which hops are applied, is the preserving malt liquors. The shoots called hop tops are introduced, in spring, as a vegetable at table, and somewhat resemble asparagus.

Reel more than a century forwards: my grandfather fed his daughters hop shoots. Reel two centuries forwards,

I fed my own children hop shoots. They do, indeed, taste like asparagus.

~

Gertrude Partridge (1868–1960), daughter of the Adams/Partridge family at Woodston, wrote about the mid-Victorian-era hop-picking on the farm:

> About hop picking. To begin with, we had 250 pickers – Black Country folk – nail makers principally and they spoke dreadfully. Staffordshire drawl! When they came, each family had a pair of sheets and a counterpane. They slept in the apple room with beam partitions [for the apples being stored], hanging hurricane lamps at intervals. Every morning we sent a big barrel of coffee out to them in the hop yard about 7.30. We made the coffee in a large furnace or copper and during that we made a large furnace of soup which we doled out to them at night. The bread came by cart from Tenbury twice a week given out two or three nights a week. On Sunday, we gave each picker a dinner, four dinners possibly on a dish which was served out to them. The dinner consisted of meat, potatoes, a slice of suet pudding, a piece of bread and some of the broth the meat was boiled in in the furnace. The women, three or four of them, came into the back kitchen on a Sunday morning and peeled the potatoes where they were boiled in three capped pots

on the fire. I don't think we ever fed more than 250 pickers; as we grew more hops and had to have more pickers, the dinners were given up and everyone was given 6d instead. Carts of vegetables and meat came and they bought what they wanted and very glad we were to be free of so much work.

~

Hops were not the only fruit at Woodston in the Victorian 'High Farming Years' (1840–70). There were also the apples, damsons, cherries for 'fruit liquor', along with Hereford cattle (as beef and dairy), some Shropshire Longhorns. Many of the cattle were stall-fed rather than free-grazing. Stall-fed cattle required, obviously, stalls, and at Woodston, as with similar Teme-side farms, a courtyard was planned out beside the house, large in ambition and grandeur. The Gentleman Farmer wished for picturesque farming, rather than peasant economy.

In the High Farming Years, the old church of St Lawrence was knocked down and rebuilt, Gothic-style, in red sandstone in 1862, the Reverend C.W. Landor contributing to the cost of rebuilding the church before donating money for the school opposite. (This single-storey stone building was attended by my mother and her four sisters.) To allow children to participate in hop-picking, local school holidays were extended into September.

~

There was an embarrassment of jobs for children down on the Victorian farm. According to the contemporary government paper 'Work Performed by Children on Farms in the 1860s: Report on the Employment of Children, Young Persons and Women in Agriculture, 1867–69', these comprised:

January – hop pole shaving and other coppice work in woodland counties

February – twitching, stone picking, bean and pea setting

March – potato setting, bird scaring, cleaning land for spring corn

April – bird scaring, weeding corn, setting potatoes

May – bird scaring, weeding corn, cleaning land for turnips, bark harvest, tending cows in the lanes, etc.

June – hay making, turnip singling

July – turnip singling, pea picking, cutting thistles, scaring birds from ripening corn

August – corn harvest, gleaning

September – hop harvest, tending sheep or pigs on the stubbles

October – potato and fruit gathering, twitching, dibbling and dropping wheat

November – bird scaring from new-sown wheat and beans, acorning

December – stone picking, spreading cow dropping; in Norfolk scaring birds from corn stacks; in Essex, helping their fathers to make surface drains; in woodland districts, coppice work; topping and tailing turnips, cleaning roots for cattle

~

By a strange synchronicity I discovered the above, in an anthology of British history, having just sent my ten-year-old daughter off to clean out the pigs. Then again, my very first paid job in farming was collecting stones, a penny a bucket. I was seven, and the Rolling Stones were on the Philips transistor radio beside me.

~

William Cobbett encountered Worcestershire hop-growers in 1826, and they were people fiercely proud of their hops.

Worcester, Tuesday, 26th Sept.

The hop-picking and bagging is over here. The crop, as in the other hop-countries, has been very great, and the quality as good as ever was known. The average price appears to be about 75s. the hundred weight. The reader (if he do not belong to a hop-country) should be told, that hop-planters, and even all their neighbours, are, as hop-ward, mad, though the most sane and reasonable people as to all other matters. They are ten times more jealous upon this score than men ever are of their wives; aye, and than they are of their mistresses, which is going a great deal farther.

The hop patriotism of the Worcester grower was as ardent a century later; the 'Fuggles' and 'Goldings' hops from Woodston went to Bass in the 1940s, and Bass only took the finest.

~

The hopyards of west Worcestershire had their own language. In the yards a caterpillar was a 'hop-dog', a moth a 'ghoop-oulud', a tendril a 'wire'. Beyond the yards, in the fields and the barns, the local farming vocabulary was as opaque as it was picturesque:

Mawkin – scarecrow

Aurrust – harvest

Wolly – rows into which hay is raked

Yarby-tec – decoction of herbs

Miale – to rain slightly

Costrel – drinking flask

Copy – small coppice

Come-back – guinea fowl

Chitterlings – entrails of animals

Trunkey – small fat pig

Carlock – charlock

Murfeys – potatoes

My grandfather, while he always retained a burr on his tongue, usually spoke the Queen's English (not least because, from time to time, he managed land for the Prudential and had to negotiate with its be-suited executives, as well as the buyers at London's Hop Exchange). He would, though, occasionally slip into conversation an arresting local idiom: 'ranald' was a fox, an 'oont' a mole, and a 'quist' a wood pigeon. Indeed, in the trove of west Worcestershire/east Herefordshire dialect, birds were particularly exotically styled, thus:

Spadguck – sparrow

Old-maid – lapwing

Catahrandtail – redstart

Eeckle – woodpecker

Hedge betty – dunnock

Cheat – grasshopper warbler

Mumruffin – long-tailed tit

Almost gone now, the old words and superstitions, though on occasion they still bubble up from the deep past. A few years back I was at Worcester market buying sheep, and the vendor was keen on detailing the finer points of his 'yoes', which he grazed in his 'archert', where he also ran 'gulls'. After an initial flummoxing, I guessed correctly at geese for the last of these.

~

There is a moment when Woodston enters national history; the exact day is unknown, but it is 1884, the event so well known locally that it was passed down via word of mouth through the generations, so that in 1978 James Downes, whose father worked for the Partridges, knew that:

> Just over Eastham Bridge at the bottom of Feather-
> bed Lane on the right hand side is two acres of land.
> [Part of Monks Meadow.] This was the first hop-
> ground to have wirework instead of poles. But it
> wouldn't work as the hop vines would not climb the
> wire. It would go about two feet then slip down.
> Hearing that in Saxony the framework was poles
> and wire and the vines climbed string and produced
> large crops of hops, Mr James Partridge went to

Saxony to investigate, and found this to be true. On arriving back home the new method was tried out and proved a great success.

The present, national, system of supporting the growing hop bine from a lattice of wirework was introduced by the farmer of Woodston, James Adams Partridge, eventually inheriting the farm in 1892. During the hop-picking season he employed 2,000 pickers.

According to James Downes:

Jim Partridge had a system of colours to denote which part of the Black Country the pickers came from. Each colour had a Head Woman who did all the negotiating with Mr Partridge. The allotted colours were painted on the ends of the crib and the work poles at the end of the rows were painted with the colour of the group who had to pick there.

At the bottom side of the Eastham bridge there was a double row of wooden barracks . . . they were for housing hop-pickers. I think there were 36 facing the road and 36 facing the river . . . Mr Partridge had another 2 ranks at the top of the Lakes hopyard besides farm buildings in which to house his 2,000 pickers.

My Amos grandparents' lives were dedicated to, and ordered by, the farming of hops. Even when Joe and

Margaret Amos retired they, as mentioned earlier, strung up decorative hops in the sitting room, the dining room and the hall, but they need not have bothered, because after a lifetime the sharp fragrance of hops had entered their clothes and, I fancy, their skin. As a boy my grandfather smoked dried hops, rolled up in pieces of newspaper; as a man he graduated to Woodbines, choosing this brand, I always suspected, because it had the hoppy word 'bine' in it.

~

Last night the sky was black velvet sprawled with diamonds, so cold this morning was a given. At around seven, the sheep rose, leaving the white frost-sheet spotted with their warm, green circles of sleep. Zeb, my horse, cantered up for its food, its hooves making a hollow drum of the paddock; I thought the sound a magnificent cavalry charge. But I'm prone to romance.

Until I begin the secretarial work that is the substance of modern farming.

At ten I am in the office, where the inside of the windows exhibit frost-etchings of Jurassic ferns. I turn on the HP laptop, but only after I turn on the two-bar electric fire.

In the bad old days filing animal movements took five minutes, a paper form filled in triplicate and posted to Herefordshire animal health department on Blackfriars Street. With the wonder of computers and rural broadband, the same process takes an hour. I can't get on the requisite website because the internet is too slow, then the page with the form freezes, so I have to reload; there is a problem with 'submission'. Restart the application.

By now the study is full of fug and fugue; it's a damp room anyway, but the frost on the windows is melting. Condensation runs down the panes like the legs of wine in a glass. Or tears.

I am startled from my repining by a tapping at the bottom corner of the window. A smudged blue tit, balled against the cold, is pecking at some crack in the putty, with desperate determination. The *pick-pick-pick* goes on for minutes. She must realize that the fracture will not yield, but such is her hunger she continues. She hopes.

I wonder why she is not at the bird table on the lawn, and swipe a pane of the glass with the sleeve of my jumper, and then understand. Perched, ungainly on the wooden roof, is a juvenile buzzard.

A bang on the glass scoots it away, and little birds flock out of the trees and bushes back to the bird table. And what do I see? A redpoll. I am certain of the identification, but check anyway with Rev. C.A. Johns' *British Birds in Their Haunts*, the 1938 edition, gifted me by my mother. Inside the cover reads: *Kathleen Amos, Woodston*.

The screen on the HP Stream informs me:

No internet. Try:
    Checking the network cables, modem, and router
    Reconnecting to Wi-Fi
    Running Windows Network Diagnostics

By now I am hot, if only under the collar, so I throw open the office window. Glass can be as much of a prison as iron bars.

The sounds of the countryside in winter surge in with the glittering cold air. The playground's squabbling of the bird table; a pheasant in our stubble patch beating its wings; further away, behind the wood, crows honking like a flock of taxis.

A van toots its horn; postwoman Patricia. I walk out to the postbox, hung on the front gate, open the jiffy bag sticking out of its gape; a package from Stephanie Mocroft, an historian, and expert on matters Lindridge. She has sent two photographs showing dried hops being bagged in the Woodston kiln; the columnar jute bags, taller than the man propping one up in photograph 1, are stencilled 'James Adams Partridge, Woodston, 1896'. With a griffon, the mythological half lion, half eagle beloved of English heraldry. The man himself is sitting beside a bag, dressed in tweed and moustaches. The equally hirsute tallyman is holding the pages of the record book open.

In the second photograph, seven workers are assembled alongside James Partridge Adams, and every one of them with a cloth cap on his head, and pride on their faces.

Make no mistake: these are Victorian trophy photographs, akin to those with great white hunters posing with foot on the dead tiger.

The beast slain/The harvest bagged.

The same thing.

The kilns themselves, built by Adams in 1880, were a statement in wood and red brick (made in the brickyard at Eastham, half a mile away) to Victorian certitude.

There are buildings that sit on the earth. Perch on it. The Woodston kilns *squat* on their blessed plot.

The hops have gone from Woodston but their monument remains. The first time I saw the square complex of kilns, with their four white wind-driven cowls, I had the wind taken out of my own sails. It was a bright spring day, but the mighty, massive gable-end kilns at Woodston are a sight on any day in any season.

And a reminder of the productive power of farmland before the chemical deluge.

To me, the kilns are another sort of memorial. My parents held their wedding reception in the bagging room.

～

Life at Woodston in the mid-Victorian century is easily pictured. The Victorian gentry were fervent letter writers and diarists, the Adams family and the Partridges they intermarried being no exception.

Gertrude Partridge writes of life at Miles Hope, another Adams/Partridge family farm in the Teme valley, but the routine was that of ivy-clad Woodston too, with its veranda under which the Adamses sat for formal familial photographs; the family moved frequently between their houses:

The cattle we had were all Herefords so we did not have a large dairy, only enough for household use and to make butter. Two large pigs were killed at a time (4 each winter) and there was much to do in putting them away. Phillips the game keeper cut them up and did the salting, Mrs Phillips cleaned the entrails in a swift running stream and used them for making pigs puddings; they were stuffed with groats and flavoured with herb organy. The groats were cooked with lumps of fat before being put in the puddings. They were tied up and a whole copper full would be boiled at a time. They had to be pricked while boiling to prevent them bursting. They were eventually hung up across the kitchen in loops and were very nice when fresh and fried. We ate them till we got tired of them or they were mouldy! No sausages were made but instead pork pies, 24 or 25 at a time and cooked in the oven. There was plenty of lard and liquor, this being a thinner grease than lard; it was put in jars and tied down with a bladder, labelled and used for medicinal dressings. The salting was done dry with a little saltpetre in most bloody places, hams were done just the same and kept turned very regularly. None of it was smoked. After curing, the flitches and hams were sewn up in calico, hams hung up and flitches laid on big racks in the big kitchen.

Washing was done once a month, two women came in to do it and were washing for two days. Sheets and table cloths etc were boiled in the furnace and the rinsing in fine weather was done out in the yard. Edith [Gertrude's little half-sister] was very fat and she fell in one of the tubs one day! My sisters and I did the ironing in a room upstairs at the back which had a big stove in the middle for heating the irons. There would be a big clothes basket full of starched collars, cuffs, dicky fronts and white shirts to polish. At Christmas time, there would be two women plucking geese in the back kitchen and Mother would be drawing them in the big kitchen. She would pull out the 'leaf' this being a lacy network of fat, lay it on the breast and decorate it with ivy leaves all the way up the middle finishing with sprigs of holly and mistletoe. These would then be sold at Tenbury market.

And still are sold at Tenbury market, and the orchards of Woodston still bubble with great balls of mistletoe. Farmer's diaries are few and far between – because few farmers have the time to put pen to paper – but one that does exist from the Victorian century, with the serendipity that should visit every historian's endeavours, is that of Peter Davis, who ran a farm almost identical to Woodston at Burford, six miles away; he was in the same social set as the Adamses – which had more than a touch of Jane Austen Regency about it – and, as it happens, he also tried to lease 'Woodson', aka Lower Woodston, aka Woodston Manor.

1836

5th April Slow rain all day. Go to Tenbury Market. Meet with Mr. James Adams and try to treat with him for Woodson Farm but he wants too much – 320£ a year rent for 189 acres besides Tithes which are high and all other bold strokes in my judgement.

April 7th Stormy, mild. William at Kidderminster Market. Oyster dinner at Swan.

8 April Solid rainy day. Another great flood. When will it cease? Such a rainy spring surely was never known before. Had another conference with Adams about the Farm but he wouldn't come low enough.

22 April Mild. Slight storms. A party of Gents dine with us from Tenbury Fair viz Wm. and Edward Mytton, J. Smith, James Adams, E. Edwards, Mr. Partridge & Geo. Winton.

2nd May Dry, some sun, but very cold wind. Go to Ludlow Fair and buy a nag of Mr. Evan Price of

Ackel near Presteign 49 for 20£. A real roadster or cob 50 about 14½ hands, colour grey.

7 May Small frost, very mild fine day. Proceed on to Worcester from New House. Take some cash and do some business at the bank. Meet with Mr. Turnall there. Dine with him along with a few more at the Hopmarket Inn. On returning home between Eardiston and Lindridge meet with a chaise containing the new married couple Mr. J. Smith & Late Miss Mytton, the ceremony having taken place at Lindridge this day. Revd. Mr. Powell, the officiating clergyman, got quite mellow as well as some of the other company.

25 May Sharp frost, clear warm day. Had a right hard day's sheep-shearing. Five of us shore nearly 157 (all large ones). Mr. Edwards not here. Mr. Partridge comes in afternoon, it being Ludlow Fair.

27 June Fair & warm. Dine with the Adams's at Eardiston having gone down to have a leisurely peep over George Adams's Farm at Upper Woodson. Mrs. Nurse and Mrs. Bright leave Park for the Lodge. Mr. & Mrs. Henry Davis & Miss Henderson from Liverpool call in the evening.

June 30th Excessively hot. Go to Ludlow Races, being the second day. Only three races to be run and for these we have 2 heats, the Oakly Park being walked over for and the handicap decided at one heat . . . Father goes to Kidderminster Market. James

Adams comes to Tea and have another word with regard to Woodson but could not treat. Asks now 300 guineas, bid the 280£ again . . . Mr. & Mrs. H. Davis & Miss Henderson call after Tea to invite us down to the concert at Swan on Thursday night.

July 1st Very warm & fine. Father and William go to pay rent to and dine with Mr. Rushout. Maria goes to spend the evening at Nash. I go to an evening party at Mr. H. Davis's. We play at cards, dance, eat supper &c. till 2 clock.

3 Aug Same weather. Go across to the Lea to see Mr. Wm. Lowe & also call on Sarah Pugh to get her to hire some hop-pickers if they be had in that neighbourhood.

11th Aug Same weather. Father goes to the Clee hill to hire hop pickers – gets 14. Calls on Mr. Bishop of Lowe to pay for the cheeses, William goes to Kidderminster Market.

Sepr 6th Stormy A.M. Fair P.M. Begin hop picking, Father and I been at it all day.

26 September Same weather. Finish hop-picking. A meeting of the guardians &c. of the Tenbury union. A fair at Tenbury & a Dinner at the Oak at which William staid – 46 sat down. After paying the hoppickers we have a dance amongst them on the green by moonlight. Robert Sayer calls and sups.

29 Nov Excessively heavy rain all night as well as day. The quantity of rain fallen of late is immense.

For several years we have not had such a flood in Teme. Reaches half way up both the Adneys and covers nearly the whole of the lower one. Mr. Wheeler's reys hopyard completely inundated and the greater part of the reys meadow. We put a fox out of the cot in orchard at Burford and the dog runs him into the river. 9 o'clock at night. Just returned from saving a lot of nine sheep from being drowned in the middle adneys, the water having risen so rapidly since the edge of night as to cover nearly the whole of the meadow. Find the sheep in the middle of the meadow surrounded by water entirely. Bring cattle and all into Spurtree meadow. Get a fall from the horse beginning to plunge whilst in the water amongst the ditches. Very vivid flashes of lightning. The deepest flood that has been for a number of years.

Feby. 8th [1837] Mild, some gentle rain. Walker & his hounds bring a fox which they had below the hawthorns – dig him out and kill the gentleman before many yards.

9 April Severe frost again, very cold east wind accompanied with storms of snow, hail &c. Sunday. Dine & spend the evening alone: go to church twice like a good boy.

The weather, the farmer's lament today as yesterday, same as it ever was.

Then again, life in the countryside has its incomparable comedies to brighten the day. I sometimes have to explain to people from cities and towns that *The Vicar of Dibley* TV series is not actually humour, more searingly real documentary. One of my current proofs is a clipping from the *Tenbury Wells Advertiser* of 17 April 1906, headlined 'Not a Painted Name':

> JA Partridge, farmer, was summoned before the Tenbury Police Court, on Tuesday, for a breach of the Highway Act, by not having his name painted on a cart in his possession. PC Butler, in proving the case, stated that he was on duty in Mamble, when he saw a coal cart belonging to Mr Partridge, in charge of a man named Morris. The name J Partridge, Woodson, was only written on the cart in chalk, and the witness told the man in charge he should report the matter. The name should have been painted on. Defendant had been previously cautioned by the police for the same cart. The magistrates inflicted a fine of 4/-.

By 1906 the money had long gone out of farming; a demonstration in Trafalgar Square against cheap foreign hops in 1908 attracted 50,000 farmers and agricultural workers. If James Adams Partridge being nabbed by the local bobby and fined for failing to paint his name on a cart has the ring of comedy, the report in the *Tenbury*

Advertiser from the same year concerning a man jailed for stealing a 4d hop-pole has the ring of desperation.

The price of wheat had crashed, crashed again. The farmers of Lindridge were somewhat shielded by their diversity of product, their mixed farming – Woodston was *circa* 20 per cent hops, 30 per cent fruit, 20 per cent arable, 30 per cent livestock grazing – but even so the Adamses needed to mortgage the farms, and jiggle finances.

Two letters from James Adams to his sister Gertrude, a mere decade apart, are eloquent in demonstrating the gilt coming off the *doré* decades:[2]

Woodston House Mar 9th 1884

My dear Gertie,

I wish you very many and very happy returns of your birthday and I hope you will forgive me for not writing yesterday, my wish was to do so but somehow the time could not be found. Many thanks for letter to me; the hounds met at Abberly Hall on Wednesday last so there was something going on as a birthday treat [Jim's birthday was March 5th], about 60 or 70 horsemen came to the Meet and a select few, self included, went in to breakfast. The inside of that place is simply gorgeous, such painting and gilding, carpets and silver; it is without exception the most elaborate display of wealth unlimited I have ever seen, you can have no

idea of what it is like, he gave us a very good cham-
pagne breakfast and so forth; afterwards we soon found
a fox and had a good run from Abberly to Ribbesford
Woods, Charley rode very well and was in front a good
deal of the way. Sir Francis and Lady Winnington
were both at the Meet. The next day Thursday the
Ludlow hounds met at Henley Court and Cast. Grea-
torex gave us a grand champagne luncheon. We found
a fox and ran it into Wire Forest. There now, I will try
and change the subject as you have had enough of
hunting I shd. think . . .

Your loving brother
J A Partridge

Woodston House Dec 29 1893

My dear Gertie,

Lindridge House and Land
You will bear in mind that I have already paid £1,547 to
clear the above property from all encumbrances, this has
been done and the deeds are at the National Provincial
Bank Leominster the place is worth about £1,400 at the
outside, and 1/7 of it belongs to each one of us.

Mother brothers and sisters occupy house gardens &
I occupy about 9 acres of land and value it at a rent of
£28 per annum. I propose paying this rent at the end of
each year, or if more convenient 1/2 yearly beginning

with a payment of £4 for each 1/7 share which I enclose for the current year of 1893 – the receipt for which to be in full discharge of all claims against me on account of the property up to this date.

You must excuse me writing this letter in business form because being business it is necessary,

With best love to you all,
Believe me,

Your loving brother
J A Partridge

Farming? It is all relative. The gold gloss of big money was removed from Woodston, but passing through Lindridge in 1902 on his cadastral survey of the countryside, a second Cobbett, H. Rider Haggard, concluded: 'here [is] beautiful land – I should say some of the best in England'.[3] A hill farmer from the Welsh borders passing through today might still eye it with envy as 'easy land'.

As I do.

~

The poets may praise the mellow mists, but what is special about October for real ruralists is the vital, exhilarating mornings, with the sky so wind-wiped clean of cloud you can see clear into the stratosphere. And then the English Holy Trinity of smells on the frost-edged air: rotting leaves, bonfires, and gunpowder from Eley cartridges.

All of these, of course, are a sort of incense.

There is only one place to be in autumn, and that is in the country, and within that geography, in the classic landscape of field and hedge, brook and copse.

For years now in this history of English farming, using my 'Method Writing'™, I have tried to live the period concerned. So, having ploughed with an antler like the Prehistorics, made 'tree hay' in the style of Saxons, scattered seed from a waist pouch as a medieval peasant would have done, eaten a Tudor and Stuart wedding cake (made from meat, surprisingly[4]), scythed hay like the Victorian 'Hodge' (that era's slang for farmworker), I have reached my Edwardian Age.

Accordingly, on this fine morning of *circa* 1904, I am out and about with a .410 shotgun, seeking something for the pot. In my conjured scenario I am a tenant farmer; consequently the targets for lunch are 'vermin' such as coney and pigeon – the high value, high falutin' 'game' belongs to my landlord.

I own to an incongruity. I have a dog beside me, a Black Labrador, but one without a single shred of retrieving ability, or indeed any work ethic whatsoever. Plum comes from a line of international champion Labradors . . . for Beauty. She can lie with her hind legs crossed in the most decorous starlet manner, but when the dummy rabbit is thrown she looks at one, and says, 'Oh, I think not.' She is, however, damn good company. The best days have dogs in them. Every day should be a dog day.

So we mooched up to the wood, but there was nothing there, except the soft shuffle of footsteps through fallen leaves, the arpeggios of laughter from the jays, and the music of light wind through the October trees in their Klimt-gold dresses. (Only trees die elegantly.) At the junction in the stone path, we took the route most travelled, down to the wheat field in the valley, halting by a hazel bush the grey squirrels had unaccountably overlooked. I picked about a pound of bonneted nuts into one of the paper bags I always have in the pockets of my Harris tweed jacket. One's farm, in 1904, was expected to provide food in every way it could – hence my morning of 'walking up' shooting.

When we reached the big flat wheat field by the brook, a grey cloud of wood pigeon rose from the severely barbered stubble before I could get within a hundred yards, and blew off over the hill. Some sable rooks stalked up and down the straw rows, looking grave, in the manner of gowned headmasters invigilating exams.

I have eaten rook, and I was contemplating raising the gun when they intuited my intent, and all the five black birds that might have made a pie lifted off, and slowly, lazily oared to a point beyond pellet-reach.

Clever birds, rooks.

The rain of last week was retained by the clay soil, so proceeding through the centre of the field was hard walking; the dog lifted her nose in displeasure. Crossing the glistening nail bed, I saw nothing except one cock pheasant, who sprinted neck forward for the winning line of the woods, a multi-coloured shooting star, but along and over the surface of earth.

More pigeons arrowed across the wide valley, the sunlight catching their breasts so they flared white, which super-charged their speed and made them unobtainable. Reaching the end of the field, where it bangs into the old village quarry, I decided to turn clockwise and beat around the top, woodland edge of the stubble field, the safety catch of the .410 off; a 'Frenchman', a red-legged partridge, sped off on its comical, whizzy little legs; I thought the morning would be a blank, until fifty yards further I disturbed a rabbit which bobtail-bolted away, but then fatally changed direction to offer its flank.

The rabbit bucked, once, twice: a little jig of death, a break-dance of extinction.

We all end in indignity; excrement and urine pouring out of us; gurgling; wide-eyed and white-eyed.

This is the truth that farmers know. Needless to say, I made the retrieve of the rabbit myself.

The rabbit suffers the injustice of all common things, in that its beauty is overlooked. But in picking up the doe rabbit, her fur — with its complex hues of bronze and grey — was soft in my hand, and I could not immediately think of anything more gentle to the touch, other than the hair on the head of a newborn baby, or a dandelion clock, or perhaps the breeze over the meadow on a June evening. Her pelt was as shiny as sunlit water.

I am often asked, 'If you are a nature lover, how can you kill things?' I take no pleasure in killing – although I do take pleasure in the skill of the hunt, and the satisfaction of bringing home the mammoth, duty done – and if I farm for wildlife, is it not right that wildlife provides me with the occasional meal?

My 1904 self has certainties I do not. He is a natural naturalist, who learned the bounty of edible October mushrooms holding his mother's hand, whose father showed him the 'smeuse' (the doorway in the hedge) of the hare. He does not suffer the guilt that the killing of beasts and birds might somehow deplete Nature. There is so much of it in 1904. The Edwardian sky runs with turtle doves, the bright brook scuttles with water voles, the woods echo with the barking of the deer, in every hopyard a covey of grey partridge may be put up, and into every cottage comes a queen buff-tailed bumble bee seeking an hibernaculum.

It is good to be out in a morning in autumn 1904, before winter comes.

~

When *Country Life* magazine commissioned a writer in 1913 to bicycle from Perranporth in Cornwall to Cromer in Norfolk for a series on rural change, what struck him was the way most farmers clung to the horse. 'If they did have a machine and it went wrong their instinct is to leave it to rust beside the hedge.'

Life in the countryside, despite the machine and capitalist patterns of employment, remained stubbornly careful and slow. Hayricks were often given time-consuming caps twined from straw into an individual pattern. Wood was still a prime material for construction, and woods on farms were still managed, as they had been in Norman times. In 1878 Woodston auctioned, at the eminently suitable venue of the Oak Hotel in Tenbury Wells, the standing coppice timber from the Ask Coppice on the farm. There were 1,410 oak trees and poles, 67 scotch fir trees, 33 ash, plus wych elm, sycamore, alder, willow, larch, beech trees, birch, poplar, Spanish chestnut, cherry trees and aspen.

In the 1960s, a retired farmworker in his eighties told R.E. Moreau, author of *The Departed Village*: 'People didn't know no different than to just keep jiggeting along; and whatever they were doing, they were interested in. People were more civilized then.'

There just seemed to be all the time in the world. In fact, time was running out. War was around the corner.

CHAPTER VIII

THE TIES THAT BIND

1914 to 1955 – The Great War – William Farmer
Pudge – My grandparents move to Woodston – The rise
of the tractor, the fall of the horse – Hop-picking in
the Second World War – Tales my mother told me –
The wild plants of Woodston – Poppop's
retirement – The end of hops – Woodston today

'The grand secret is to match the breed to
the soil and the climate.'

William Youatt, Agriculturalist (1776–1847)

THERE IS A PHOTOGRAPH, black and white, of my grandparents in 1938, standing with three of their daughters,
Kathleen, Margaret, Daphne; Madeleine and Josephine,
the twins, have yet to be born. It is eight years to the day
since they arrived at Woodston, the day my grandfather
took over as farm manager.

The photograph says everything you need to know:
Poppop, dressed in farmer tweed and flat cap, is typically

confident, happy, looking you straight in the eyes. Touch-ingly, he has his hands on daughter Marg's shoulders; he is an affectionate man, and idolized by his girls. My grandmother looks wary, as well she might. A war with Germany is on the cards, despite the Prime Minister's promise of 'peace in our time' after that Munich meeting with the Führer; the house, a brick cottage, which comes with the job, is too small.

Then there is my grandfather's new employer, William Farmer Pudge. Farmer Pudge, who, amusingly, does tend towards stoutness – as well as engagement in agriculture – in a true case of nominative determinism, owns, aside from Woodston's 100 acres of hops, 400 other acres of hopyards round about, plus thousands of acres of other land. He is one of the most successful farmers of his generation.[1] On his death in the 1970s, his daughter will inherit £49 million, causing my grandparents to say in unison, 'He never made that from farming'; they were right, he invested well and wisely, and Woodston is the best hopland there is. He has got Joe Amos in to run things, despite his relative youth (he is still in his twenties). The problem with the arrangement is Pudge's estate manager, with whom my grandfather does not see eye-to-eye.

My grandfather is a proud man, and not keen on suffering fools lightly. By the time of the photograph, my mother Kathleen, who is ten, has been to eight schools, the moves more than not occasioned by my grandfather arguing with the owner. However, he always finds employment because he is, as all concur, an outstanding farm manager. Having left village school at thirteen, he struggles to read the *Daily Express* but has a keen intelligence and a near faultless memory. He can reel off to my grandmother all the men's hours and wage rates; she, in a very traditional division of labour on the farm, does the book-keeping.

He knows everything there is to know about farming; he knows everything there is to know about Nature. He is also one of those rare men able to command, without offence, largely because he knows what he wants, and when. He is briskly efficient, as befits a man nicknamed the Whippet. At times the Woodston workforce will rise from the usual twenty to nine hundred as the seasonal hop-pickers come in, mostly from the West Midlands, along with a hundred or so Romanies in their painted caravans.

I suspect there are more reasons for my grandmother's face of doubts in that 1938 photo. Farming is down in one of its familiar doldrums, hops especially. It is also a time of bewildering change down on the farm.

~

The golden age of the hop industry was the nineteenth century. Hop acreage continued to increase until 1878 when it reached its peak with 77,000 acres. Tastes changed: there was a decline in the demand for hoppy porter and a surging demand for a lighter beer known as Indian ale or pale ale. Pasteurization arrived in the late 1870s and fewer hops were needed as preservative. There were only 32,000 acres of land growing hops by 1909 and a renewed import of foreign hops, because breweries contracted to brew foreign beers were required by law to use the hops stipulated in the original recipe. (The popularity of lager would give the industry another kicking in the 1970s.) Twenty-three years later and acreage had fallen to 16,500. My grandfather, you might say, was lucky to get the job he loved.

Then there were the machines. In 1897 a Mr Locke-King bought a newfangled vehicle from the Hornsby firm in Lincolnshire – the first recorded sale of a tractor in Britain. The kingdom's farms would never be the same again. For two glorious centuries the Shire horse, the Suffolk Punch and, in Scotland, the Clydesdale, had strode magnificently over the farming landscape of Britain, from valley to hilltop. The Great War had been no help to the horses; they had been rounded up by army impressment gangs and sent to serve abroad, 160,000 of them in August 1914 alone. The German U-boat campaign, which targeted supply ships en route to Britain, then brought the country close to starvation and it became a case of grow more food at home or die. The obvious answer was to bring more land into cultivation. Since the farm horses had been co-opted as war horses, tractors had to be imported from America to take their stead.

Numerous small tractor companies were formed, but all struggled to compete with the Model F and Model N tractors of Henry Ford. The 'Fordson' was produced at Dagenham in England in 1933. Thereafter, blue and orange Fordsons became indelible sights in the British countryside of the 1930s. A Fordson in 1935 cost £150 for the basic model, £195 with low-pressure pneumatic tyres. The Fordson won a place in the heart of the farmer, congenitally keen on saving money. A US government test in the 1930s concluded that farmers spent $0.95 per acre ploughing with a Fordson, whereas feeding eight horses for a year and paying two drivers cost $1.46 per acre. Tractors, as the early advertisements gleefully pointed out, only 'ate' when hungry. A tractor put in the barn for the night consumed nothing; a horse had its head in a nosebag or hay rack for hours. By 1939 there were 50,000 tractors on British farms.

The end of the horse was bewilderingly swift. In 1914 there were 25 million horses in the UK; in 1940 there were 5 million, of which only 600,000 were on farms.

There was hardship for those who lived by the horse. In fact, down on the farm in Britain in the 1930s there was hardship all round. A serious decline in agriculture from 1922 was hitting farmers' profits (by 1934 home-grown wheat had reached its lowest market price, accounting for inflation, since records began in 1646), while electrification and mechanization, both of which reduced labour costs, were making the labourer redundant. Whereas in 1926,

200 farms had electricity, in 1936, 6,500 did. One penny-worth of electricity fed into a milking machine would milk 40 cows – a man's work for a day. And a man cost more than a penny a day to run, so the man had to go. By 1930 only 12 per cent of the British workforce was in agriculture; once again, the smallest proportion for any country in the world.

Disease, naturally, continued to rear its ugly head. One English farmer wrote in his blue leather-bound *Country Gentleman's Diary* in 1935: 'All our 4 cows passed TT [tuberculin] test – Maria – Dolly – Beauty – Carole.' Aside from listing his purchases of cattle and equipment, it was the only event this farmer thought worthy of recording in his diary, save for major bust-ups in the House of Commons. Herds were small, animals loved.

If you were a farmer or farm manager there were thirteen notifiable diseases, outbreaks of which needed to be notified to the police and local authority by the terms of the Diseases of Animals Acts, 1894–1922. These included cattle plague, foot and mouth, anthrax, bovine tuberculosis, rabies and swine fever. Sheep-dipping orders, formulated to control sheep scab, laid down that all sheep in England be dipped once in the summer, and in Scotland up to three times. In some scheduled areas, sheep had to be 'double dipped', that is dipped and then dipped again at a period not before seven but not after fourteen days. The process was overseen by the local policeman, who bicycled to the farm.

The owner of a sick animal had a number of options. Nature could be left to take its course, the (pricey) vet could be called out, or one could administer any number of proprietary cures, all widely available, all widely advertised. *The Country Gentleman's Estate Book & Diary* – my grandfather's bible – extolled 'Tippers Cow Relief, made at The Veterinary Chemical Works, Birmingham. SAVES THE UDDER – your source of income.' The Cataline Company of Bristol, in the same venerable publication, claimed that its drench 'For All Chill and Inflammatory Udder Trouble' was 'Unsurpassed'. Six bottles of the medicine would treat twenty-four cows. It was also a balm for all farm animals, and could be dosed to sheep and pigs at one-eighth of a bottle per head.

Quacks – unregulated vets – visited farms almost weekly trying to sell their potions. If the expensive vet was called out, while he was on the premises, Poppop tended to make good use of him: at Woodston, my mother had an abscessed tooth removed by the vet, but only after he had tended a sick Shorthorn heifer; the farm had a small milking herd of ten, the milk left in aluminium urns on the roadside for the milkman to fetch. (The Shorthorn was yet to be supplanted by the Friesian.)

Under the plough-up campaign of 1914–18, two and a half million acres of ancient pastureland, including 50 acres of Woodston, went under steel blades; hares, my grandfather noted, were not as numerous in the local countryside at the end of the war as at its start (he was born in 1904; shooting hares for the pot was a teenage task, equipped with a single barrel .410 shotgun). There had also been a great loss of farmland birdlife during the war, by order of the government, who targeted birds they identified as grain-eating pests. Previously protected species such as house sparrows were now categorized as 'vermin' which could be destroyed by poison. (Shades of the Tudor Vermin Acts.) Soldiers in training in England were sent off on official bird's-nest destroying expeditions. 'Shoot the birds', urged the *Daily Mail*.

It was all change down on the farm post-1918; food prices were low; the housing stock insufficient, and rents tended to be more than agricultural workers could afford. Farmers complained of a shortage of workers, which led to more mechanization, fewer horses, fewer people.

The Partridge scion James Adams Pennell de Woodstone Partridge had been killed in the First World War, aged 23 and serving with the Canadian Army,[2] and the heart was knocked out of the family. Death and taxes caused them, like many landowners, to sell up; the aristocracy and the gentry, contra the fashionable wisdom that they shirked battle and the 'working class' bore the brunt of slaughter, perished in disproportionate numbers because they were the officer class, and officers were first over the top, last to retreat. In the decade following the end of the war, about a quarter of England changed hands, with farmers the major buyers; by 1927 36 per cent of farmland in England was owned by its farmers, compared with 12 per cent before the war. The Woodston estate was divided into bits, the Upper and Lower farms going separate ways, and then into more bits.

In St Lawrence's church, Lindridge, there is a memorial to James Adams Pennell de Woodstone Partridge. I do not know why he decided to volunteer in the Great War, but I like to think he was one of the young gentlemen, like the poet Edward Thomas, he of 'Adlestrop' fame, who fought for King and Countryside, as much as King and Country. When asked by his metropolitan friends why he was going to the frontline, Thomas picked up a handful of English country earth, and said, 'Literally, for this.'

~

On Tuesday, 16 November 1926, Woodston Manor Estate was auctioned at the Swan Hotel, Tenbury Wells, in lots. The sales particulars read: 'Over 491 Acres comprising some of the Finest Fruit and Hop Land and Fattening Pastures in Worcestershire'. Woodston Manor was 'A very Charming House', with nine bed and drawing rooms, came with a bailiff's and a gardener's house, modern stabling, farm buildings and 87 acres of land with cherry, damson, plum and pear orchards in 'full bearing'. Sold in another lot was, 'One of the finest hopyards in the Midlands with modern hop kilns and machinery, range of cattle yards and two cottages'. The remainder of the sale was 'fattening pastures, several small fruit holdings, orchards, accommodation land, woodland, and a mile of trout and grayling fishing in the River Teme'.

Woodston was an apple, divided many times.

In 1930 when my grandfather moved to the 'bailiff's house', Woodston Manor comprised 100 acres; Upper Woodston had reverted to independence, but it was the newly created 'Woodston Farm' (duly entered on Ordnance Survey maps as such) of 150 acres, including the 'modern hop kilns' and most of the associated hopyards, that he managed. There was no farmhouse as such; there was my grandfather and grandmother in Mill Cottage running the farm for the absentee owner, William Farmer Pudge.

Despite the quartering some things remained the same; elements of life at Woodston (whether Upper, Manor or Farm) in the 1930s would have been familiar to Peter Davis, even the Penells of the seventeenth century. Chickens scratched on a muddy yard. The rural year was still based on an ancient calendar of spring lambs, summer shearing, summer hay, late summer wheat and autumn calves. Cattle were generally 'in wintered' (put in barns) from October to May, to be fed hay or straw, or specially grown forage fodder such as mangels, swedes or kale. The milch cows dictated a daily rhythm of milking at 5.30 a.m. and 4.15 p.m. The major jobs on the farm were mass communal enterprises. When Woodston Farm's flock of fifty Kerry sheep required shearing, ten men sat on benches, hand-shearing, sorting, carrying, putting on the owner's marking with pitch, dabbing picric acid on the cuts on the sheep's body. The people helping were fellow farmers, their sons and daughters, villagers in want of a day's work. It was a community event.

This is the world of Woodston my grandparents entered. The permanent labour force at Woodston Farm, in addition to the farm manager and his wife (and his five daughters), consisted of a ploughman, a cattleman, a stable boy, two milkmaids and four general hands.

~

Why, you wonder, given farming's 1930s woes, did my grandfather want to farm Woodston? Well, there was the money, obviously. There was the challenge of managing the best hop farm in the country, the variety of the work; aside from the hops there were the cattle (about thirty Herefords for beef, plus the Shorthorns for milk), a few sheep (mostly to vary the grazing, and so improve the sward), a variety of arable crops, and a few cider, cherry, plum and perry orchards, making the scene a mass of white confetti in spring. (The 1923 Census of Fruit Trees recorded a total of 1,834,621 fruit trees in Worcestershire, made up of, inter alia, 766,100 apples, 136,327 pears, 53,065 cherries and 857,144 plums; most but not all of Woodston's fruit orchards, in the great post-war division, went to the parts not farmed by my grandfather, leaving my mother and her sisters having to 'scrump' cherries on the way to catch the bus to Kidderminster, on a Saturday morning expedition.) There were also the Landrace and Gloucester Old Spot pigs,[3] much cossetted by my auntie Marg.

It was a genuinely mixed farm.

More than this: he loved being outside, and he loved the artistry of a fine furrow (ploughed with a Fordson, then a Ferguson). He had no fear of the machines, though he kept a couple of Shires for when the tractor broke down, as it did. He loved seeing the land bear fruit.

And Woodston was beautiful. He said it was the most beautifully located of all the farms he worked (though 'the kilns were a bit of an eyesore').

And farming is what he did. It was him, and he was it.

~

The countryside of the 1930s was still quaint; the distinctive hedgerow 'quilt' was intact, even in the arable areas of the east, and would remain so until the 1950s. It was still the land of 1914 that John Masefield memorialized:

> *How still this quiet cornfield is to-night!*
> *By an intenser glow the evening falls,*
> *Bringing, not darkness, but a deeper light;*
> *Among the stocks a partridge covey calls.*
> *The windows glitter on the distant hill;*
> *Beyond the hedge the sheep-bells in the fold*
> *Stumble on sudden music and are still;*
> *The forlorn pinewoods droop above the wold.*
> *An endless quiet valley reaches out*

Past the blue hills into the evening sky;
Over the stubble, cawing goes a rout
Of rooks from harvest, flagging as they fly.
So beautiful it is, I never saw
So great a beauty on these English fields,
Touched by the twilight's coming into awe,
Ripe to the soul and rich with summer's yields.
These homes, this valley spread below me here,
The rooks, the tilted stacks, the beasts in pen,
Have been the heartfelt things, past-speaking dear
To unknown generations of dead men,
Who, century after century, held these farms . . .

Weekend tourists came to Woodston to view the famous hops beside the Teme, some in small saloon cars, others on bicycles. My grandmother gave them a glass of milk; she was always keen to chat.

~

Hop-picking tales my mother told me: she saw raven-haired gypsies dance wildly around fires; a knife-fight to settle a blood feud; she ate baked hedgehog with a gypsy boy called David; a hop-picker from Stourbridge hung himself in the kiln. Until the Second World War, the gypsies arrived in painted wooden caravans towed by horses; by the end of the 1940s they came in aluminium caravans pulled by lorries.

Every September the little world of Woodston became

a city, as the Romanies and the pickers from the West Midlands, Birmingham, Dudley and Stourbridge arrived on the train at Newnham to walk the half mile to the farm. (The Tenbury & Bewdley Railway, 'The Blue Bell Line', which brought the pickers, also took the Amos girls to school in Kidderminster, and my grandfather to the Hop Exchange in London; it closed in 1964.) On the farm, the pickers from the West Midlands lived in the wooden 'barracks', a shed per family, glad to be out of the city with its grime and Nazi bombs. Almost every day the ice-cream man visited the yards on his motorbike and sidecar (the sidecar was the 'fridge'), as did the local butcher, vicar, Methodist minister, grocer and fishmonger, and once a week a film was shown, projected on a screen in one of the barns.

The hop-picking day was nineteen hours long, organized into two shifts from 6 a.m. to 3.30 p.m., and 3.30 p.m. to 1 a.m. to get the hops picked and dried. Dinners were cooked over iron grates called hop devils (barbecues are not new), breakfast and lunch tended to be makeshift, and sometimes most of the working day was maintained by 'dripping cakes', a penny a piece from the perambulating baker. Workers were now paid in money; as late as the 1920s they had been paid in tokens to be spent at the local shop, or cashed in at the end of their work, or what they called their 'holiday', because it was nicer than city life. (As George Orwell witnessed, London's East Enders similarly treated hop-picking in Kent as a rural vacation.[4]) The money for their wages was collected by my grandfather every Friday from the Lloyds Bank on 71 Teme Street, Tenbury Wells, as cash, and driven to Woodston, where it was dispensed by my grandmother. The village pubs, of which there were three within staggering distance, did good business on a Saturday night.

All the picking at Woodston was done by hand, even though hop-picking machines were under manufacture down the road in Suckley and Malvern, because my grandfather was convinced, after watching the machine at work, that it massacred the delicate hop 'cone' or flower; the hops, pulled down from their wires, were 'picked' (stripped) into canvas cradles. The women all wore floral shirts or dresses, or, if it were damp, gaberdine macs. The prelude to picking was my grandfather walking every jungly hopyard, up and down the alleys between the towering green tresses, taking a sample cone, rubbing it between thumb and forefinger to determine whether the golden resin, crucial in bittering and preserving, was ripe. It was on his call, and his call only, when to pick, in which yard and in what order. Get it wrong, either too early or too late, and the crop could be worthless, something no brewer would pay for.

When the hop-pickers returned home in September, Poppop drove to Kidderminster in a lorry to collect shoddy, the unwanted wool from clothes manufacturers; this was one of the main manures for the hops at Woodston. It was all organic in those days.

Norman Jones, from Halesowen in the West Midlands, picked at Woodston in the 1930s and 1940s:

I'd got an Austin 7 by that time and I'd made a trailer for fastening on the back. When I got the word . . . that the hops were ready, I would set off in the Austin 7 with my wife and daughter, with the trailer on the back piled up with our luggage. I'd also got a belle tent that we pitched on a little green patch. We went to Woodston every year from 1930 – I don't think we ever missed – until we stopped going a few years after the war.

I met a gypsy named Aaron Locke down there . . . While we were there hop-picking, Aaron died. They buried him in Lindridge cemetery, the most beautiful cemetery in Worcestershire, and all the hop-pickers went to his funeral because he was well-liked. After the funeral, I witnessed an amazing thing. Aaron's children made a pile of straw in the field and set fire to it. When it was well alight, Aaron's eldest son, Ivor, dragged the caravan over the fire and burnt it. They said it was a Romany tradition to burn a gypsy-man's house when he died . . .

Jones was a nailmaker, Halesowen a town firmly within the Black Country, the sooty industrialized West Midlands. Whole Black Country families would uproot for the hop-picking of Worcestershire and Herefordshire, a six-week holiday with pay (1s a day for a good picker as early as 1846), and, perhaps even better, for fresh air and the natural beauty of it all. In Jones's words, they all went home afterwards 'brown and with full bellies'. In 1908 2,500 men, women and children left Halesowen, and neighbouring Cradley and Old Hall, for the annual Nailmakers' Holiday, carrying their 'hoppen baskets', containing the kit and caboodle needed for the vacation.

The Nailmakers' Holiday was dead by the early 1960s, withered on the bine by the mechanization of the hop-yards, and the demise of the nail-making industry in the West Midlands, the men gone to factories and mines indisposed to grant a six-week jolly out in the green, the schools persecuting parents who took their offspring away from the rote-learning of term time.

The end of annual, mass hop-picking was just another severance between country and town, the making of a nation divided as much by geography as class.

～

Other hop-picking and Woodston tales my mother told me. During the Second World War the Land Girls came in 1944, and one got pregnant and had to be sent away; the pickers were not allowed outside fires at night between 1939 and 1945 in case these lit the way for Luft-waffe raids on South Wales; the worst days of the week were when the 'slop man' came and emptied bins of waste food from hospitals for the pigs; the farm received a Ministry of Information leaflet on picking rosehips for the Vitamin C, and she and some other girls from Lindridge were dragooned by the vicar into picking rosehips for a

Vitamin C tonic, and her hands were so scratched she could not hold a knife and fork properly afterwards. (The leaflet was 'Hedgerow Harvest'; by 1943 around 450 tons of rosehips were gathered annually.) Then again, picking and sorting hops tore the hands 'to ribbons'.

Once there was a sign on the Peacock Inn saying 'No Beer', which was plain funny in a hop-growing land. My grandfather worried over the scarcity of lapwings, which were great birds for eating pests; he supposed it was because villagers were taking their eggs, to supplement their rations. My mother and her sister Marg were allowed to go by train to visit their aunt Barbara, who lived in Hereford, twenty-five miles away. There were people in Lindridge who considered Hereford 'foreign'. People did not travel far. A trip to the Regal Cinema in Tenbury Wells, which had opened in 1937, was 'a treat'. A somewhat exotic one indeed, with its Mediterranean murals by artist George Legge painted around the auditorium.

There was also the boy refugee from London, who stayed in the main house. Later he wrote a poem about his sojourn, 'The Visit, 1939':

> *'I've not the strength' he said,*
> *that waistcoat-suited man,*
> *short, round, with*
> *thinning hair*
> *and gingerish moustache.*
> *From the city.*

'I've not the strength' Dad said
to Mother
as they plodded in the
hot autumnal sun
from Newnham Bridge to Lindridge
(no taxi, for the war was on
and petrol was in short supply)
to visit me,
evacuated
to a heartless
manor house.

But they were well-received
(contempt of County for the Town
disguised)
and given tea and cakes.
I know, because on that
Singular occasion
I was admitted to the living room,
the inner sanctum of the house
that was forbidden territory
to us evacuees —
except when parents visited
(they innocent of this otherwise
unviolable law).

And when they left that house
I walked beside them,

past endless rows of hops
in fields that lay
on either side the road
to Newnham Bridge
me, pleading to go home with them,
they, torn between my pleading
and their fear
of bombs falling on our house,
pleading with me
to stay,
my waistcoat-suited father
stopping, breathless,
ev'ry hundred yards or so,
to walk within the limits
imposed
by his failing heart,
my mother looking
anxiously at him
and tearfully at me.

'I've not the strength' he said
But when we parted,
he had the strength
to weep.

The boy grew up to be the biologist Sir Gabriel Horn, Master of Sidney Sussex College, Cambridge. Despite the acidic notes of the poem (published in Sidney Sussex

College's Yearbook), he came to adore the Teme Valley and revisited numerous times before his death in 2012.

~

My mother's most enduring memory of wartime Woodston – as it was for most farming folk – was the visit from the 'War Ag Committee', who came to check the farm's efficiency because, once again, in time of crisis, the farmer and the land were asked to produce more. The job of the War Ag Committees was to improve farms and bring uncultivated land under the plough. Woodston Farm's official inspection came in 1943; the official departed happy in his Austin car with a bottle of cider and a ham in the well of the passenger side. In this 'Second Domesday Book' my grandfather, as manager of Woodston Farm, passed 'A', the only official recommendations being the ploughing up of one meadow (which Poppop thought a shame, because 'to break a pasture makes a man; to make a pasture breaks a man'; flora-rich pasture takes an age to make) and more drainage for certain fields above the A443 (probably the perennially wet Stephens Moor and those neighbouring it), though the snipe, and the other birds of wet 'marginal' farmland, may have derided the Man from the Ministry's drainage advice.

~

Lover of swamps
The quagmire over grown

283

With hassock tufts of sedge – where fear encamps
Around thy home alone

The trembling grass
Quakes from the human foot
Nor bears the weight of man to let him pass
Where thou alone and mute

Sittest at rest
In safety neath the clump
Of hugh flag forrest that thy haunts invest
Or some old sallow stump

Thriving on seams
That tiney island swell
Just hilling from the mud and rancid streams
Suiting thy nature well

For here thy bill
Suited by wisdom good
Of rude unseemly length doth delve and drill
The gelid mass for food

And here mayhap
When summer suns hath drest
The moors rude desolate and spungy lap
May hide thy mystic nest . . .

From 'To the Snipe', John Clare

~

The grand national programme of wartime ploughing up brought results that widely exceeded the war government's wildest expectations. In 1943 there was a record harvest; the season's target of new acres to be ploughed had been set at 960,000 but farmers managed 1,376,000. In the process the mechanization of British farming proved unstoppable. Tractors and farm machinery had improved beyond measure, and the demand for them raged. In spite of the priority given to military vehicle construction, the number of tractors in Britain rose from 56,000 in 1939 to 203,000 in 1945. The number of disc harrows doubled between 1942 and 1946. Farm incomes shot up between 1938 and 1943 by 207 per cent, compared with a national average of 35 per cent.

~

Is there anything lovelier than an English country lane in May? I'm walking the lane down through Woodston, my daughter's Jack Russell with me. This is his afternoon constitutional. He has age, twelve years of it, and he has SARDS (Sudden Acquired Retinal Degeneration Syndrome).

Wholly blind, he operates by nose and by ears. He is lost in smell and sound. I, sighted, am lost in the colours of the verge flowers. I am as absorbed as any Romantic poet. So, Snoopy and I wander along together, cloud slow, behind our senses, me on the tarmac, him pushing through the sward with the determination of a tug in high seas.

Overhead swallows chatter in air already easy with the promise of summer; the hedge is heaped with hawthorn blossom; the chaffinch by the gate has laid her four eggs, painted with runes; orange-tip butterflies dance.

Normally all these things would divert me, but we are in mid-May when the wayside flowers fountain from the earth. When they tone perfectly with each other and the grass itself.

Inlaid in the green border of the lane there is the pink of campion, the blue of forget-me-not, and the white of greater stitchwort. I bend down and pick a small sprig of stitchwort: the white of stitchwort is pure white. Bridal white.

A scene like this, twenty years ago, inspired my wife and I to have the wildflowers of the wayside made in icing to adorn our wedding cake. The decorative flowers were made by a mistress cake-maker up at Michaelchurch Escley, a raven's soar over a hill and a mountain from Llanthony church, where we tied the knot.

We still have the wedding-cake flowers.

Clearly, I am in romantic as well as Romantic mood.

The little dog pushes through a towering clump of cow parsley, and a host of black, tardy St Mark's flies rise and drone away, their heavy legs hanging down.

Today even cow parsley seems graceful, worthy of the folk name Queen Anne's lace.

As Snoopy and I amble, the list of verge flowers grows. Strewn before us is sow thistle, ribwort plantain, small-flowered cranesbill, herb robert, yellow creeping buttercup and speedwell, whose bright blue flower was once sewn by travellers into the lining of their coats as a charm: 'speed-well'.

There is a green wave of cleavers (with unexpected, star-tiny flowers) crashing into the hedge. The dog clambers over the cleaver tendrils, to the easier going of ground ivy, shy and creeping. Except: Snoopy's favourite smell is mammal scent, and ground ivy is plant menthol. He sniffs, then puts on a surge of speed to cause a bow wave in the ground elder.

A laneside verge is so much more than a bit of grass; a six-foot swathe such as this is old-time meadow, and an unacknowledged Nature Reserve. The charity Plantlife has counted 720 species of grass and flora on Britain's verges.

Is there any strip of land more useful than the verge? Perhaps, but not many. Cottagers in Worcestershire used to let the house cow loose on the lane for free foddering. On the drovers' roads into the county from Wales the fat grassy margins were grazing for beasts on their way to London's meat markets. All before the coming of the car, of course.

The verge was also a source of free food for the rural dweller. Well, it still is, on this lane. Thus, the unromantic me has in one hand a Sainsbury's carrier bag to fill with wayside foragings. May's wildflower days are also salad days.

It occurs to me, as I pick a spire of garlic mustard, that botany is a form of historical record, a living archaeology. In the wild plants and flowers that survive there is continuity, connection with the practices of our rural past. My past. So the dog and I walk up the footpath from Woodston to Lindridge Church, checking the flora as we go, and we find, *inter alia*:

Blackberry – part of the British diet since Neolithic times.

Blackthorn – the fruit makes sloe gin, a favourite tipple of my grandfather before going hunting.

Burdock – known to my grandparents by the local name of gipsy's rhubarb; used in the drink dandelion and burdock.

Chickweed – young poultry are said to thrive on this common annual. So too pigs, who grow speedily on its high protein (15–20 per cent) and iron content. We feed it to ours.

Coltsfoot – the herb's botanical name comes from *tussis*, meaning 'cough-dispeller'. My grandmother gave me some concoction containing coltsfoot when I was a child. (It was as retching as the spoonful of cod liver oil I was doused with each evening before bed.)

Broad-leaved dock – old and venerable treatment for soothing nettle stings on children's skin.

Ground elder – in days of yore, ground elder was held to be good for relieving the symptoms of gout. Since ecclesiasts were believed to be chronic sufferers of the disease (on account of their taste for wine and rich food), the folk name of bishop's weed followed as a matter of course. Likely used by the monks of Lindridge.

Ground ivy – three of ground ivy's local names, alehoof, gill-over-the-ground (from the French *guiller*, to ferment) and tunhoof, are relics of its role in the brewing industry from Saxon times until its supersession by hops in the sixteenth century.

Hawthorn – until the 1950s, young children walking to school in the countryside would munch new hawthorn leaves and shoots as a wayside nibble, called in many places bread-and-cheese. These included my mother, possibly picking from this self-same bush. The hawthorn is also reputed to divine bad weather. 'Many haws, many snaws,' runs the old wife's tale, here meaning the wife of Joe Amos.

Sorrel – a corruption of the Old French *surele*, meaning 'sour'. Until the time of Henry VIII, sorrel was cultivated as a herb and used principally in a 'green sauce' for fish. Sorrel's Latin name means, roughly, 'suck vinegar', because agricultural workers sucked the leaves to slake their thirst. Myself included on July afternoons.

Elder – the blossom for elderflower cordial, the berries for elderflower wine. John Evelyn thought that an extract of the berries was a 'catholicon against all infirmities whatever'. In historical times, the 'Englishman's grape' was grown commercially in orchards, Worcestershire included.

Hazel – or 'filbert' locally. The culinary use of hazelnuts is prehistoric, and they formed an important item in the diet of Mesolithic hunter-gatherers. Hazel sticks are the rods used for dowsing water. My grandfather could dowse for water.

Mistletoe – popular to deck the halls from the seventeenth century, and for a chaste kiss. Tenbury Wells is still the capital of mistletoe sales, the town with the last remaining mistletoe auctions, these held by Nick Champion in the old Round Market building. Mistletoe likes fresh air and apple trees, so it likes orchardy Woodston and west Worcestershire. Almost all of it nowadays is picked by Roma people. The Druids' beloved plant has an association with fertility.

Legend says that the Norse goddess Frigga made every plant and animal promise not to harm her handsome son, Baldr, the god of vegetation. But she overlooked the mistletoe. Taking advantage of this, the capricious sprite Loki made a mistletoe spear and tricked the blind Höðr into killing Baldr with it, thus bringing winter into the world. Luckily, the gods restored Baldr to life, and in celebration Frigga declared that two people passing under mistletoe should kiss to celebrate Baldr's resurrection.

The little dog and I end up in Lindridge churchyard, where there is a swathe of old grassland flowers, among them lady's bedstraw and agrimony, and slow-growing lichens that enjoy the clean air. We potter around to the front for the view, avoiding the porch (the masonry is falling down). The long stepway in front of the church was climbed by my parents in 1952, when they married here. There's a photograph of the wedding procession in which a formidable hatted aunt stops halfway to admire the view, above this meander of the Teme. Below her a policeman, wearing white gloves, directs the parking of the cars.

～

1956, six years after my parents married at St Lawrence's, Lindridge, is the year my grandparents left Woodston Farm. Hops were bottoming, and only the recently created and producer-controlled Hop Marketing Board prevented the extinction of the industry in Britain; by 1960 there were a mere 16,000 acres of Britain down to hops. (The Hop Marketing Board continued to function until 1982 when the EEC decreed it was illegal, being a monopoly.) The last hop-pickers from the Birmingham area had come to Woodston Farm in 1954; men could no longer get the time off work, and women with school-age children could not take their children out of school.

Pudge, a keen businessman, sold up, and Woodston Farm ceased to exist, its acres and its hopyards – including the two acres where James Adams developed the poles-and-wire system of hop-growing – going to neighbours, among them the pre-existing Woodston Manor and Upper Woodston. Once again, the kaleidoscope of Woodston parts were shaken into a new pattern.

My grandparents in that year moved to Woodend, at Cradley, which was hop-less, and after contracting sciatica Poppop retired. He was never averse to modernization, though spraying 'chemicals' troubled him, and he failed to see the point of ripping up hedges, since they 'kept the soil from shifting about'. By the 1970s, 5,000 miles of hedges were being 'lost' per annum. He also wondered where all the cuckoos had gone. Together we heard what was probably the last corncrake in Herefordshire; this was at Withington, about 1972. He marvelled at nitrogen fertilizer until the truth became apparent: it reduced the diversity of the sward; 'a lot of growth, not much goodness'.

Poppop retired in 1962 . . . for about a year. At Withington in Herefordshire he took on a part-time job as farm bailiff, chiefly managing hops. I can see him now, returning from the yard in his brown smock coat and black Wellingtons, clutching a mass of hop bines for my grandmother to 'dress' their house with. I don't think I have ever seen anyone look happier.

~

Going through my mother's effects I found a photograph of my grandfather ploughing; he is in the 'saddle' of a Ferguson TEF 20 tractor, bright and shiny new; it is raining. I know this because his mac is as sheeny as the Ferguson's tinwork, and the peak of his tweed cap is pulled down, in the way you do when you want to keep the rain out of your eyes. There is no 'cab' on the tractor.

My guess is that the photograph was taken *circa* 1955, the Woodston days drawing to a close. It is no exaggeration to say that this photograph haunted me, although it took me an eternity, then a revelation, to understand why.

The Damascene moment occurred in a field in west Herefordshire in the summer of 1999. I had spent hours in the cab of a tractor, radio on, air-con on, going up and down a 25-acre field 'topping' – pulling a rotary mower to raze weeds.

As I exited the field, I realized that I had not set foot in it, breathed it, touched it. I knew from my position behind 5 mm safety glass (I might as well have been sitting in a skyscraper office down Canary Wharf, to be honest) that it was sunny, but the actual temperature? I had no idea. Smells? Ditto. Bird song? Ditto. What birds, beasties and bugs were down in the grass? No idea. An elevated bubble.

I drove home, then took the doors off the cab of the International tractor for a literal breath of fresh air. That evening, I dug out the photograph of my grandfather ploughing at Woodston, then saw the, literally, blinding obvious: the health of the soil, the way it gleamed with vitality. His deep concentrated connection with it all.

My grandfather's 'Little Grey Fergie' was pulling, no problem at all, a three-furrow plough through clay soil. I do not wish to get technical, scientific, even disputatious, but these days, due to soil compaction and deterioration, you need a tractor the size of a house and the price of a house to get a three-furrow plough through much of Britain's agricultural ground. Then, that same machinery compacts the soil into a 'pan', and you need an even bigger tractor to pull the plough.

I was not exactly the insensitive, agri-business type before that eureka moment in 1999, but then I decided to fully embrace the future – which was my grandfather's era, the past. Organic. Treading lightly on the soil.

Understanding, appreciating, that Nature and the farmer should be a union

This is one of my personal inheritances from Woodston. As sort of proof, my current main tractor is a Ferguson TEF 20, built in 1956. An old tractor, farming the old way.

~

A disquisition on Fergusons, and the independence of farmers.

My Ferguson diesel tractor is now over 65 years old, and was built in the year of the photograph of Poppop ploughing. (Who knows, RWF 321 may even have been his Ferguson: the square number plate on the mudguard in the photo is blanked black by dirt.) Is my Ferguson retiring? Is it hell. Fergusons go on, and on.

The Ferguson, made at the Standard Motor Company plant in Coventry between 1946 and 1956, is one of farming's true icons. Small yet butch. When Sir Edmund Hillary decided to cross the Antarctic he chose a caravanserai of 'Little Grey Fergies' for the job.

Admittedly, a vintage or classic tractor – meaning, essentially, anything built before 1980 – will not win a drag race against a banana-yellow JCB Fastrac or an ice-white Lamborghini Nitro 100 VRT. Ploughing at 3 mph on a Ferguson is life in second gear, in the slow furrow. But is that a bad thing in a world spinning madly?

Since there is no cab on the Ferguson, I get to look

around, see things. The comet-burst of starlings over the stubble. The fluttering kestrel over the sheep paddock.

I get to smell things, above all, the ferrous odour of fresh-turned earth.

I get to feel the turns of the year; the fieldfare-frost of winter, the butterfly-sunlight of summer.

My grandfather once said, unwittingly, a truth about his 1939 Fordson, which applies to all tractors of vintage make: 'She was a good ride.' One sits in a modern tractor; you *ride* an old tractor. You are in the saddle, bucking up and down. One has the best of both worlds with an old tractor: animal horsiness and mechanical hp.

A contemporary tractor can easily cost £100,000, and you pray that the electronics never go wrong. And you lie awake at night, counting not sheep but repayments to the bank. (Small wonder the suicide rate in agriculture is disproportionate.)[5] Such tractors, with their enclosed cabs and computer software, isolate the farmer from his/her husbandry; the experience of land-tending becomes mediated absolutely by technology, the farmer reduced to remote, robotic overseer of food production.

We farmers like to consider ourselves the last independent heroes, but we have become slaves to banks, government subsidy, machinery, supermarkets, the job itself.

~

Another revelation, this time about livestock, of which my grandfather would approve . . .

Katabasis. The ditch was about 8 feet deep – you get a lot of rain on the hills of the Welsh borders – a shadow world of ferns, moss, slime, serpent ivy, slinking water. And at the bottom a ewe, broken, puppet-sprawled. She needed killing, humanely. I did it myself, rather than wait hours for the vet. I put the 12-bore against the back of her head, pulled the trigger. The blast detached her head, so it was attached to her body by a thin string of marbly white skin only. Blood, strangely scarlet and fluorescent, seeped slowly into the water.

I believed I had distracted the rest of the flock with a bucket of sheep nuts. Not so. In the moment of execution I realized that, curious, they were on the edge of the ditch looking at me.

They then ran in fear to the far end of the field.

Sometimes the hardest thing in life is to acknowledge reality. I had not exactly been a hard-hearted farmer before that moment but, on clambering out of the ditch, I was obliged to drop my cognitive dissonance, my objectification of sheep. In a kaleidoscopic moment I saw that the flock was not a monolithic unit but composed of sub-groups based on friendship and family bonds. One old ewe, Sooty, had movingly gathered her daughter and granddaughter about her.

The flock did not come near me for weeks. But then sheep have excellent memories, and remember faces – ovine and human – for years.

Maybe I should file that incident in the Black Mountains as my Damascene moment, rather than my trip into the Hellenic underworld. As a result of my sheep's reaction I realized that I wanted a relationship of compassionate companionship with my flock. My views have since evolved into principles of meat-rearing and meat-eating ('ethical carnivorism') that can be briefly stated thus:

We should eat less meat.

The meat we do eat should be sustainable and Nature-friendly.

Farm livestock are not Cartesian flesh-robots but sentient creatures deserving of a good life and a good death – which should come only near the end of a natural lifespan.

Yes, I hold attitudes one would expect from a shrub-drinking hipster outta East London rather than a fifty-something farmer from the west of England, who has raised livestock for twenty-five years, and whose family have farmed for eight hundred.

Unsurprisingly, I also have a beef with the intensive end of my industry, with its beak-clipping, tail-docking, permanent 'in-housing', zero-grazing, nitrogen-spewing, Frankenstein cattle-making, prophylactic antibiotic-dosing. Raising of livestock in this fashion is not farming, because it abjures any sense of husbandry. It is senseless, inhumane Fordian food-production of 'units'. Also, the produce from such factory systems, be it milk, meat or eggs, is tasteless, in every sense.

So back to the future, to the old-fashioned husbandry of my grandfather, and the farming of traditional breeds, able to thrive without shiploads of imported soya or the vet's constant administrations.[6]

Back to old-fashioned cooking, as well. My grand-parents ate the animal from tongue to tail. Not a part was wasted. Bones went to soup, brains to 'rissoles'. Animals' lives are too valuable for our current 'prime-cut only' wastefulness.

~

Coming to terms with change . . .

A travel article in 1967 said of Lindridge: 'Across the road a farmer still tends his herd but the hop kilns next to the church no longer dry hops, they have been converted into someone's home.'

I cannot provide provenance for the piece; it was a clipping found in my mother's effects when she died in 2010. (It was inside a book, sheltered; she wanted to die in good order.) I cannot provide explanation, either, for her keeping of the piece, except to suggest nostalgia. It was 1967 that she first took me to Woodston.

Each place has its own special smell, and that of Woodston in spring 1967 when I toddled alongside my mother was softly intoxicating floral: the fruit trees were in full white blossom, and the orchards on the banks – to my infant mind – seemingly steeped in snow. Of course, my mother may simply have wished to collect paper portents. Woodston's own kilns, after years of echoey dereliction, were converted into apartments in the early 1980s. My mother was intrigued, and, in alliance with her sister Daphne, arranged a viewing. She came away befuddled by how 'swish' the apartments were, and the utter loss inside of any trace of the place's original role. Significantly, the kilns had been renamed 'Woodston Oast House' by the developers, who presumably considered 'oast' (a Kentish word) a more charming cognomen than the local and workmanlike 'kiln', and thus all the better to attract incomers.

After my mother's nosy reconnaissance, she showed me a photograph of her wedding reception from 1952, which was held in the low, beamy bagging room of the kilns, with everyone sat at long tables covered with white linen, the men in suits or tweed jackets, the boys in blazers from public or grammar school, the women in hats; a girl clutching a posy. The pillars of the bagging room had been carefully woven with foliage.

If the photograph required a caption it would be: 'A rural wedding reception, 1952.'

Something else, and it comes off the photo more blindingly than the photographer's flash on the champagne bottles: the sense of homogeneous country community. 'Incomer' was an unborn word.

~

I went nightwalking yesterday, nightwalking being a habit begun as a teenager, because the congenial village pub that served Wadsworth's 6X to the underage was not, alas, situated in my village, so required a three-mile perambulation home at closing time.

To walk the British countryside at night is to enter a dark, adventurous continent.

At night, the normal rules of Nature do not apply. In the night-wood I have met a badger coming the other way, tipped my cap, said hello. The animals do not expect us humans to be abroad in the dark, which is their time, when the world still belongs to them.

That was in winter. The screaming of a tawny owl echoed off the bare trees. For all of our streetlamp civilization, you can still hear the call of the wild if you go out after the decline of the day, which is the human's safe time.

The nightwalker has an alerted sense of hearing. Our ears receive the notes and tones of the countryside undetected by purely diurnal beings, and their cars.

The hops have gone from Woodston, and Lindridge, but there are still some hopyards at neighbouring Newnham. So in the November wind I walked along the frosting road, the moonlight blazing on the Teme, notes of iron from the clay soil in the air, the stars my guide, and kept my ears open.

Then I heard it, the sound of wind in the hop-yard wires; a pastoral symphony for lyres.

It was a ghostly sound.

Night is for imagining, for dreaming.

~

This winter morning I went walking at Woodston, because if you need to know the truth of land, you do it in winter, when the sights are not pretty, and the wind and the cold have stripped off the disguising leaves, and the sheep have eaten down the herbage to the earth's surface.

On such a black-and-white day, you can see the bare body of the earth.

So, I stood up on Broombank, looking south over Woodston, or perhaps more accurately, Upper Woodston, Woodston Manor, and the kilns — the area around still marked on Ordnance Survey maps as Woodston Farm.

I do not wish to romanticize 'Woodston'. The hops have gone, as have most of the orchards, but . . .

The farmer of Woodston's historic core is now Mary Walker, a mere 87, who has cared for a hundred acres of Woodston for fifty years. Her Woodston Manor farms soft fruit, Cox and Russet apples (for juice), cereals, pedigree Angus cattle (the herd founded in 1999), and Berrichon sheep, hardy and suited to low-input grassland. It remains a traditional mixed farm and, due to its enrolment in the Higher Level Stewardship environmental scheme, rich in the wildlife and flora of farmland.

If the Penells, the Adamses or my grandfather could

look down they would recognize Woodston, because the scene remains the same: small scale, intimately balanced between the needs of Agriculture and Nature. A perfect English farm.

A state of mind.

EPILOGUE

In front of the Ferguson's bonnet, a sea of red furrows. Behind the tractor, the harrow makes a wake of fine tilth.

You can tell the season by the way the sun shines on clay. Winter's rays are thin milk. On a fine spring afternoon like this, the sunlight gleams buttery on the furrows.

It is life in second gear on a Ferguson TEF 20. Life at 3 mph, pulling a disc harrow up and down the field to make a seed bed. The Ferguson is 65 years old, but it goes on, and on.

Putta-putta-putta.

There is no cab. Smoke from the exhaust gets in my nose.

Putta-putta-putta.

Harrowing, unlike the geometric exactitude required by ploughing, is not demanding on the brain. There is nothing to do but sit and stare at Nature.

By the gateway, the house martins land at a puddle, grab beakfuls of mud for their masonry; the puddle is their brick factory.

Overhead, other hirondelles, swallows, are still migrating. Swallows never take the direct line but slip tipsily, left, right, in the general direction of magnetic north.

In the thorn hedge the blackbird sits hard and tight on her immaculate bowl-nest. Her home is half open to the elements; the hedge is only half leaved. She has five blue eggs. Every avian species takes a reproductive gamble; the blackbird's eggs, because they are exposed to light, will develop quickly. That is the positive. The negative is that they are easy for every petty predator in the 'hood to see.

Turning the tractor at the field's end, I look down behind at the harrowed ground. A large purple worm has been brought writhing to the surface. Blind and tender flesh, the earthworm is the original agricultural implement, plough and harrow combined. Day and night the earthworm works. Humans have never invented anything to match it.

On the tussocky margin two young rabbits squat sunning, like clods I have missed.

Suddenly, a memory so close and vivid it is akin to looking down a telescope through time. A September afternoon in a hopyard. My grandfather, fifty years a hop farmer, pulls at the cones, the hop flowers, then rubs them between thumb and forefinger to deduce if they are ripe for harvest. The yard is enclosed by high hedges, and the green hops run up wires into the sky. The air is thick with their bitter perfume. Peacock butterflies float from weed to weed. (What is a weed? A wildflower by any other name.) The men and women of Kent are right to call a hop field a 'garden'. A dreamy secluded summer garden. Shangri-La in England.

Another memory: the same hopyard, but in the black-and-white of November. I'm helping head-scarved women pull down the dead bines from the wires, which line the sky like strings on a giant's abandoned dulcimer. The bines are put on a constant bonfire; lunch is a potato baked in the ashes. A covey of grey partridge scuttle between the wooden pillars. There is beauty in this scene too, albeit of minimalist type.

Putta-putta-putta.

By now the Ferguson will be running low on diesel, so I turn it off, and walk over the furrows to the jerry can in the barn. The little journey is oddly physically demanding; in the days when men and women worked fields on foot no one was fat, except clergy and kings. In the ditch behind the barn a thousand gauzy wings vibrate. The rising tide of insect noise signs spring as surely as the coming of the swallows.

The Ferguson refuelled, I restart harrowing. Oxymoronically, I am happy harrowing, an emotional state which, according to scientists at the University of Bristol, is enhanced by the very soil itself. A specific soil bacterium, *Mycobacterium vaccae*, activates a set of serotonin-releasing neurons in the dorsal raphe nucleus of the brain, the same ones targeted by Prozac. You can get a very effective dose of *Mycobacterium vaccae* by leaning over the seat of a Ferguson.

Putta-putta-putta.

It turns out that hops are the bines that tie. On some of this harrowed land, I will grow a row of hops for personal home-brewing; a very mini-Woodston. You see, I need to grow hops, to stand once again in a hopyard, that magical secret, scented garden. Woodston is, to borrow a phrase of Ernest Hemingway's, 'a moveable feast', as well as a mindset.

Putta-putta-putta.

The sun continues to shine on me. Contra Chaucer and his 'shoures soote', April is one of the driest months. I roll back my shirtsleeves to begin the yearly tanning process known as 'farmer's arms', browned up to the elbow only.

A passing car pulls on to the verge to the gateway to make way for a livestock lorry on the lane. A boy looks out briefly from the car's window, then back to his phone.

There was only a man harrowing clods on a vintage tractor, same as his grandfather.

NOTES

Chapter I

1. *Pleut ça change*, as it were: this local news item shows a coach at Lindridge half filled with flood water in 2021: https://www.gloucestershirelive.co.uk/news/regional-news/gallery/pictures-severe-flooding-herefordshire-worcestershire-3858835

2. These Neolithics, it is now generally accepted from the DNA evidence, moved up from southern France and maybe Iberia to northern France, where they then mixed slightly with the central European population, before moving into Britain. Agriculture was introduced to Britain by incoming Continental farmers.

Chapter II

1. See Oliver Rackham, *The History of the Countryside*, chapter 6.

2. Sheep, like us, are tribal; they like to remain within the familiar tribal area, the ghetto; for them, however, the territory is indicated by scent emitted from the hooves rather than lines on maps, fences, signs by roadsides.

3. https://unherd.com/2020/08/why-mushrooms-are-magic

M. Pleszczyńska, M. K. Lemieszek, M. Siwulski, A. Wiater, W. Rzeski, J. Szczodrak, '*Fomitopsis betulina* (formerly *Piptoporus betulinus*): the Iceman's polypore fungus with modern biotechnological potential', *World J. Microbiol. Biotechnol.*, 2017.

4. In the 1920s German zoo directors Lutz and Heinz in a 'breed-back' experiment almost recreated the auroch in so-called Heck cattle. The Nazis loved them; Göring had them wandering his estate.

5. Ard marks have been identified as early as the Neolithic, for example, at the South Street Long Barrow, Avebury.

6. https://unherd.com/2020/12/stop-planting-more-trees/

I like trees. Sometimes I come over all Prince Charlesy and talk to ours, even pat their trunks. I have managed woods, written books praising trees, and I practise a bit of 'agroforestry', the farming system which combines trees with grass for grazing by Ermintrude, Shaun the Sheep, and Little Red Hen. But tree-planting in the UK is now a destructive mania. We need a moratorium on trees.

Trees have inexplicably, unaccountably, become the magic wooden bullet for all environmental ills. Anxious about flooding? Well, forget dredging rivers or digging up the suburban concrete drive, just plant a tree. Anxious about the climate effect of your cheap plane ticket to Thailand? No worries, pay a bit extra to get some minion down on Earth to stick a tree in a hole. Trying to win a general election? Commit your manifesto to trees, trees, trees. Oh Jeremy Corbyn at his last tilt at Number 10 pledged that a Labour government would plant two billion – *billion* – trees by 2040. Or, half of Wales, planted up at commercial density.

The electorate read this as 'magic money trees' and shied in the polling booth, though Boris's government, itself not lacking in *amour aboreal* (or indeed in the discovery of theurgic trees that fruit GBP), has committed itself to planting a whopping 75,000 acres of trees annually until 2025. The Green Recovery Challenge Fund – effectively an arm of DEFRA – last week allocated almost £40m to 68 projects to plant more than 800,000 trees, including 10,000 trees at 50 NHS sites and 12 'tiny forests' the size of a badminton court in urban areas.

Everyone is at it, this tree-planting palliative. I do mean everyone. The Committee on Climate Change (CCC) has decreed that we in the UK need to stick into the ground 90–120 million trees a year between now and 2050 to achieve carbon neutrality. Craft beer company Brewdog has grabbed 2,000 acres of Scottish highland to plant a million trees. The National Trust is on the band wagon; Britain's largest private landowner has promised to insert 20 million trees into its land over the next decade. The Woodland Trust, meanwhile, has launched an Emergency Tree Plan. Not enough trees yet? Danish clothing magnate Anders Holch Povlsen and his wife Anne, who own over 200,000 acres of Scotland, are removing sheep and deer across their estates to allow more native woodland.

Aye, and there's the first rub. Quite aside from the Povlsens' self-entitled decision about how to use their land (H.G. Wells' Dr Thoreau would have enjoyed their *droit de dicatator*), the Danish duo's re-treeing is removing a chunk of land from food production. Sheep and deer are rather good at providing meals from the uplands for us poor humans. Haunch of venison, anybody? Kebab? Num num.

But tried eating oak leaves? Fir cones?

It all starts to add up, this public and private tree-planting. What remains is the simple arithmetic of farming, because humans, damn them, will eat: abstract land from food production, and you are left with either a lot of food miles from importing enough to keep your population alive, or the intensification of agriculture – with all the associated pesticides, herbicides, fungicides, molluscicides – on the remaining land. It gets worse. If we are not getting protein from an Aberdeen Angus cow up a glen, it is likely we are getting it from a soya bean grown in the Amazon on what was once primary rainforest.

Make no mistake as you read this: the land is already under demand to produce more, more, and then more food. As long as a decade ago, the UN Food and Agriculture Organization warned that the global population would increase by 34% by 2050, to 9.1 billion. It added that 'in order to feed this larger, more urban and richer population, food production . . . must increase by 70%'.

So, the trees planted in order to save us from climate change . . . will kill us by starving us.

The sanctification of trees is curious. They are hailed as carbon sequesters, but so is the grassland on which cows mosey and munch. (The Intergovernmental Panel on Climate Change, by the way, had deduced that methane from Britain's ruminants is *not* causing further global warming.) Trees are lauded as increasers of biodiversity, the lament of our British naturalists being that we only have 13% tree cover and the European Union has 40%. (Continental Europe is always lovely and good isn't it?) But have these naturalists ever been to the Black Forest? Poland? The Landes region of France? The lot of them, vast tracts of silence and desolation.

I spend part of my year next door to a 5,000-acre forest in France. The parts of it *le forestier* are able to manage – by chopping and pruning trees to let in the light – are wonderful for wildlife, and nightingales sing there, but the larger unworked regions are dense, dark and conquered by invasive bramble.

So, there is the second rub. Treeland is not intrinsically better for nature. To be cruelly honest, my neighbour Philippe's chemical corn fields have more wildlife than most of our neighbouring natural forest.

Undeterred by such freely available trans-European evidence, the Rewilding Britain pressure group reported this week that it wants native woodland to regenerate through the natural dispersal of seeds. Unfortunately, without a management plan – and I looked in the small print in vain for such – this is a recipe for disaster.

I know. It goes against the grain, but if we do have to have trees for biodiversity they need to be managed, not left to their own darkening devices. Trees are hegemonic. Unchecked, trees would take over the world and place it under their shadow.

Now for the next and third rub, prompted by the clan chief on the Isle of Skye, Hugh MacLeod, deciding recently he needed to go wild and reforest 572 acres of his Scottish outcrop (that sense of territorial entitlement again) courtesy of a handy £1 million grant from the EU and Holyrood. According to chief MacLeod the island's lack of tree cover is 'not natural'.

Neither, I suggest, is living in a castle. If you want 'natural', Hugh Magnus MacLeod of MacLeod, try a cave as your abode.

Such reforesting is the thin end of the re-wilding wedge. MacLeod has already signalled his intention to reintroduce beaver. The wolf and lynx will follow, and Skye will become a rewilded theme park, with the food – no longer produced locally, of course, because all the shepherds will have been transformed into tourist guides – coming over the sea on a bonny, smoky, diesel-engined boat.

Britain is an agricultural country. It has been farmed since the Neolithic period, and today's open landscape would be absolutely recognisable to an Anglo-Saxon. These isles' celebrated 'quilt-pattern' of fields and hedges is the work of people, the result of agri-*culture*, and is not to be lightly thrown aside, or forested over. The farmed countryside is 'the past speaking dear' as the Poet Laureate John Masefield once put it. It is heritage.

Take the Lake District, beloved of that other great pastoral poet, Wordsworth. The hills of Cumbria were revealed into glory from their scrubby oak and rowan shroud by the farmer's iron axe and the gnashing teeth of his/her sheep. And well done them.

Really, why vaunt 'natural'? Some of the best places for the wild things are actually human made – meaning, the farmed environment. A traditional hay meadow may easily contain 30 plant species per square metre. The overwhelming majority of ponds in the British countryside – and there are few things better for the bugs, birds and beasties than a mere – were dug by the farmers of yore. Now *there's* an idea for mass government funding: restoring Britain's ponds.

That rasping sound? Me grinding my farmer's axe. The carbon-sequestering benefits of planting trees are frequently hyped yet rarely weighed against what existed there before – such as peaty moorland burnt regularly to renew the heather. Peatland is a particularly efficacious carbon sink.

In fact, the tree-planting proposals put forward by some conservation organisations will actually annihilate precious ecosystems. The reforesting map drawn up by Friends of the Earth and Terra Sulis marks out as suitable land the rough pastures of the North Pennines and Yorkshire Dales. Er, these farmed and thus awful 'artificial' habitats are critically important for the iconic, red-listed curlew, lapwing, grey partridge and black grouse.

Not much of a deal, is it – if, instead of the cry of the curlew, we have the hiss of wind through the needles of a Sitka spruce?

Trees are lovely, but to propose them as the one and only eco-solution is false premise. And false promise.

7. In Butser Ancient Farm trials, emmer yield when expressed as seed yield ratio has reached 1:34, and often considerably higher.

8. Ralph Vaughan Williams' 'The Lark Ascending' was written in August 1914 as Vaughan Williams walked Margate cliffs watching naval vessels on manoeuvres. With the help of the English violinist Marie Hall, Vaughan Williams re-scored the piece for solo violin and orchestra, which better still suggested the way the lark's song spreads from its high source, filling the air with a progression of fine, tinkling notes. It is this version which has entered the collective head and heart of classical-music lovers. In the Classic FM annual Hall of Fame poll, 'The Lark Ascending' was placed first in 2014, 2015, 2016 and 2017.

Chapter III

1. A calculation of total manure available at the Bignor villa on the basis of the livestock accommodation, checked against granary capacity, ox-stalls and the natural boundaries of the estate, suggested an approximate dunging of 28 hundredweight per acre annually.

2. Mutton was less favoured than beef or pork as the diet of the Roman forces in the province, but sheep bones were better represented at Barr Hill, held by a Syrian unit.

3. See Julia Best, Mike Feider and Jacqueline Pitt, *PAST: The newsletter of the Prehistoric Society,* no. 84, autumn 2010:

> The domesticated chicken has a genealogy as complicated as the Windsors, stretching back 7,000 to 10,000 years. The chicken's wild progenitor is the red junglefowl, *Gallus gallus*, according to a theory advanced by Charles Darwin and recently confirmed by DNA analysis. But *G. gallus* is not the sole progenitor of the modern chicken: scientists have identified three closely related species that might have bred with the red junglefowl.
>
> The chicken was, in all probability, introduced to Britain by the Iron Age people, though its numbers were given a

boost by the Romans, not just for the fowl's use as sport, but also for its place on the plate.

Chickens were a delicacy among the Romans, whose culinary innovations included the omelette and the practice of stuffing birds for cooking. Farmers began developing methods to fatten the birds – some used wheat bread soaked in wine, others a mixture of cumin seeds, barley and lizard fat. Out of concern for moral decay and the pursuit of epicureanism in the Roman Republic, a law in 161 BC limited chicken consumption to one per meal and only if the bird had not been overfed. The practical Roman cooks soon discovered that castrating roosters caused them to fatten on their own, and thus was born the capon. The chicken's status in Europe appears to have diminished with the collapse of Rome, and chickens reverted to their Iron Age size. As the centuries went by, hardier fowls such as geese and partridge began to adorn medieval tables.

4. 'The Ruin', from *The Earliest English Poems*, Michael Alexander (ed.), Penguin, 1991.

5. Gertie Partridge (1868–1960), offspring of the family who owned Woodston during its grand Victorian years, wrote in her dotage of her regrets: 'Above all, Woodston having to go . . .'

Chapter IV

1. The injunction from the Church of Rome against the keeping of slaves was observed in the breach as well as in the honouring.

2. In 1163 the tax caused the villein tenants of Wenlock priory to 'throw down their ploughshares' and to cease tilling the priory's lands.

3. Lindridge was well served by its clergy's ability to wrest favours from monarchs; in 1207, at the request of the prior, King John granted Lindridge, including Wodeston, the great liberties of 'sac and soc, toll and team'. The meaning is controverted but they seem to refer, respectively, to the following: 'sac and soc' to 'jurisdictional rights over certain persons or profits of justice in certain places'; 'toll', as in modern English, the right to collect 'toll' from passage of goods, animals and people (and, conversely, be free of such tariffs); 'team', 'vouching to warranty or the right to collect fees for it on one's own land or elsewhere'.

4. The thirteenth-century round is also more formally known as 'The Reading Rota' because the oldest manuscript copy was found at Reading Abbey. The line 'Bulluc sterteth, bucke verteth' means that the bullock has indeed 'started up', risen from his grassy grazing, yet he is also announcing his reproductive desires and territorial designs. The Latin word 'stertere' means 'to snore' and is etymologically a cousin to 'snorting'. The buck deer's 'ferteth' is usually translated as the sitting-room polite 'cavorting', but it is as it sounds: farteth. The deer's spring/summer diet is rich and plentiful and makes the animal, a ruminant, spectacularly gaseous. The Medieval English were an earthy race. By asking the cuckoo never to stop singing, the singer is hoping that, if the bird stays, the fair days – the days of warmth and crop growth – will stay too. Equally, the singer is likely hoping for a long life for himself, which gives a very human, very transcendent, edge to the plea to the cuckoo 'to never stop now'.

Chapter VI

1. Some parts of Lindridge were enclosed as late as 1822: http://assets.cambridge.org/97805218/27713/index/9780521827713_index.pdf

2. Cranston also built Tenbury Wells' Pump Rooms, it being believed that the waters from the well had healing properties and that Tenbury might evolve as another of England's famous spa towns.

3. William Pitt (Board of Agriculture), *General View of the Agriculture of the County of Worcester*, Sherwood, Neely & Jones, 1813, p. 277.

Chapter VII

1. By the twenty-first century, vaccination was an alternative to the slaughter of the animals to prevent the spread of such diseases; unfortunately it was not universally invoked by politicians and academics. My remembrance of foot-and-mouth 2001, first published by UnHerd, follows:

> Quiz of the week, and your starter for 10: Identify the pandemic from the following information: Professor Neil Ferguson of Imperial College computer-models apocalyptic mortality rates, the Government bungles containment of the disease, movement (and civil rights) are restricted, a miracle-cure vaccine is preferred, the economy takes a hit of billions.
>
> Covid-19? Those of us who live in the countryside might answer differently. We might reply, 'The foot-and-mouth epidemic of 2001', the 20th anniversary of which we commemorate this month.
>
> I do mean commemorate, as you do with disasters. On 19 February 2001, Craig Kirby, the statutory attendant vet at Cheale Meats abattoir in Little Warley, Essex, noticed blisters on a batch of lethargic pigs. The last epidemic of foot-and-mouth (FMD) had taken place in 1967, before Kirby was born, but he identified the symptoms correctly and called the Ministry of Agriculture, Fisheries and Food (MAFF) – the forerunner of DEFRA. Four days later, an FMD case was confirmed on a run-down pig farm at Heddon-on-the-Wall, Northumberland, which MAFF eventually determined as the source of the epizootic. (Owner Bobby Waugh was later convicted of having failed to inform the authorities of a notifiable disease, and feeding his pigs 'untreated waste'; he was

probably, to use a *bon mot*, scapegoated by MAFF, and the disease had been existent in sheep flocks for weeks before-hand but undetected.)

By the time the epidemic, caused by the Pan-Asian O variant of FMD, was brought under complete control in January 2002, over seven million cows, sheep, goats, pigs – essentially, all animal types with the Devil's cloven-hooves – had been slaughtered. Businesses had gone under, ancient rights of way closed, parts of the country deemed no-go exclusion zones. The Cheltenham Literary Festival cancelled. The General Election postponed – for the first time since the Second World War – and the economy black-holed by £10bn, from revenue lost and the compensation given to the 2,000 affected farmers.

And life down on the farm was never the same again. The fun went out of it. Farming became so micro-managed by DEFRA after FMD you could need permission to move live-stock to an adjacent field. In triplicate. While waiting six days. If you think some of Boris Johnson's Covid-19 measures smack of Big State I can only tell you that the farming community have been the guinea pigs.

We farmed sheep back then in 2001, in west Herefordshire, where England runs into Mid-Wales. I remember foot-and-mouth well, although, curiously, not as cinema, but as single-frame images. The fracturing stress of it all, I suppose. We chained and padlocked all our gates to prevent anyone, and especially the MAFF death squads, coming onto the prop-erty. One midsummer's morning I stepped out into the field behind the house, and on every ringing hill there were sky-blackening pyres of animals being burned, the barbecue stench filling the air. The village school closed due to the 'smog' from roasting animal flesh. Troops in a Land Rover, rifles sticking out the front window, going up and down the lane, the soldiers having been conscripted for the slaughter of the animals. We were under siege from our own side.

Other fragments of memory: a friend hiding her pedigree pigs in the cellar. The wheels and the underside of the stock trailer being sprayed with disinfectant at Hereford cattle market by men in those plastic suits forensic pathologists wear. Disinfectant everywhere. Disinfectant on rubber mats for cars to drive over. Disinfectant on trays for boots to step in. Disinfectant coming out of the nozzle of a spray-gun.

We lived in disinfectant. And fear.

And we who lived in the country in 2001 lost our respect for computer-types. The establishment's response to FMD was a bonfire of the sanities, not just in the science of retrospect, but in the common sense of the moment. So, let us fight through the smoke, to find the fire-starters. Beginning with Imperial College, London.

In the concrete towers of Imperial College computer-types, including Neil Ferguson, had been busy – busy modelling human diseases. At the press of a button in 2001, they decided their modelling worked for FMD, and FMD was so potentially catastrophic that the only answer was a 'contiguous cull' – the slaughter of all susceptible animals within 3km of known cases. On the basis of Imperial's computer model was established the FMD-eradication policy of Tony Blair's Labour government.

On March 16, with FMD cases at 240, MAFF announced the implementation of Imperial's contiguous cull. Enter the MAFF slaughter squads, killing millions of animals – whether healthy or not – with a bullet to the head. If the creatures were lucky. Some animals got bludgeoned. Or drowned.

The poor bloody animals, about which no-one in authority seemed to give a shit, from the RSPCA to the NEC of the ruling Labour Party. Can you imagine the fuss if a fox had been clubbed? But fine, kill lambs any which way you want to ensure their silence.

The cull was absolute, inflexible. Sheep quarantined in sheds with full bio-security? Killed. Pet sheep, disease-free, brought into the sitting room for safety? Killed. It happened to Carolyn Hoffe in Scotland, whose house was broken into by MAFF vets and their armed Gurkha escort. In Devon vets were accompanied by police in riot gear, as the former broke into farms to perform the state-approved rites of the contiguous cull – the legality of which was always dubious, whether under UK law, or EU law. Or, indeed, natural justice.

By late Spring 32,000 animals a day were being exterminated. During the summer, the figure reached 92,000, and still FMD rampaged up and down the country. In some localities, the livestock were butchered in such gargantuan numbers that the corpses lay around in mounds for days, before burial or incineration. Lorries, leaking fluid from slaughtered infected cattle, went through uninfected areas, risking the spread of the disease. Death and smoke and blood. The British countryside turned into an infernal Dantean vision in the summer of 2001. Tourism – surprise! – nosedived.

During the Vietnam War, a hapless US army officer explained infamously, 'We had to destroy the village in order to save it.' Imperial College's computer-modelling geniuses, one feels, took his advice to heart, assuming they had hearts. And 15% of Britain's farm animals were slaughtered in the contiguous cull.

Because, in the mad, mad world of FMD 2001, guess who was not allowed to control the campaign against the epizootic? Veterinary experts. They were shoved aside by an unholy alliance of Imperial College computer geeks, Blairite politicos, and the National Farmers Union, the NFU (which soon came to stand for 'No Fucking Use' down our way; the NFU is always touted as 'Agriculture's Voice', when it represents a mere 30% of farmers, and is heavily bent towards agribusiness).

In the very earliest days of the crisis, the government's very own foot-and-mouth experts – at the Animal Health Institute's laboratory at Pirbright – informed Nick Brown, the

Minister of Agriculture, that Imperial's model was flawed. For starters, the model failed to take account of the variety of farming practices, the varying rates of transmission between different species, and exaggerated the effect of windborne spread. Pirbright's Dr Paul Kitching condemned mass culling as 'scientifically unsound'.

You see, there was always an alternative to the literal over-kill of the overlooked animals in the great British barbaric barbecue. Vaccination.

Many tried to make Tony Blair see reason, including Prince Charles, who sent No 10 a scientific paper from Edinburgh University arguing the case for mass vaccination of livestock. It fell on deaf ears. Blair refused pathetically, pusillanimously to gainsay his own appointed FMD tsars, Professor David King (a *chemist* for god's sake) and Professor Roy Anderson of – you guessed it – Imperial College's computer-modelling department, the very people who deemed the contiguous cull essential. Blair was also swayed by Ben Gill, chairman of the NFU, who pleaded that vaccination would undermine Britain's FMD-free status, damaging meat exports abroad. (If you want to play the numbers game, tourism to the countryside brings in £10bn per annum – quite a tranche of it going to the many small farmers doing B&B – far more quids to the national coffers than lamb, pork and beef exports abroad; and, anyway, why not 'Eat British', Mr Ben?)

So, sense and compassion went up in smoke, along with millions of animals and billions of pounds sterling, and Britain's green and pleasant land was turned into killing fields.

It need not have happened. The likelihood of vaccination being successful was not theoretical. The Netherlands also had an outbreak of FMD in 2001. So, your second question in quiz of the week: How was the Netherland's FMD outbreak brought under control?

Correct! Mass vaccination.

2. Bridget Sudworth, *Gertrude: Yeoman Farmer's Daughter*, privately printed, 2016.

3. H. Rider Haggard, *Rural England*, Longmans, Green & Company, 1902, pp. 103–4.

4. Or rather a version of 'Bride's Pie', following the 1685 recipe in Robert May's *The Accomplisht Cook*. Ingredients in this savoury pie included 'oysters, lamb testicles, pine kernels, and cocks' combs'. I forwent the compartment filled with live birds or snakes for the guests to 'pass away the time'.

Chapter VIII

1. Private William Farmer Pudge was later a member of 'Jacob Patrol' of the Auxiliary Units, a secret resistance network of highly trained volunteers prepared to be Britain's last-ditch line of defence during the Second World War. They operated in a network of cells from hidden underground bases around the UK. Jacob Patrol were based in Bromyard.

2. He died on 20 January 1918 and is buried at Villers Station Cemetery, Villers-au-Bois, France.

3. The British pig was improved by importation of the Neapolitan pigs, so-called because of their numbers in the neighbourhood of the great Italian port of Naples; in fact, the stock *Sus indicus* originated in the Far East. Neapolitan pigs, with their docility and prodigious ability to fatten, were the basis of the productive breeds and vast bacon industries of Britain, Denmark and the Low Countries. The most important improver of the pig was the Keighley weaver Joseph Tuley, who bred one great boar named Samson, from whom every modern white pig of note, from the American Hampshire to the Danish Landrace, is descended. A good pig of 1850 would not have looked out of place in a livestock market of Tenbury Wells in the 1930s.

4. Orwell picked at a farm near Maidstone in 1931, during his 'down and out' period. He left after three weeks, exhausted by the work, hands blackened by hop resin (although his main bug-bear about the actual picking of hops was plant lice crawling around his neck). If agricultural exploitation in the hop yards dismayed him – notoriously, picked hops were unreliably measured by foremen – he was enthused by the sense of pickers' community, the relative freedom of open-air life, and the paradisical beauty of the English countryside. It is no accident that the positive interludes in Orwell's later dystopian/naturalistic fiction almost always occur in rural settings: Dorothy's hop-picking escapade in *A Clergyman's Daughter* (where, despite the gruelling labour, she is 'happy, with unreasonable happiness'); Gordon and Rosemary's tryst on Farnham Common in *Keep the Aspidistra Flying,* itself echoed by Julia and Winston's love-making in the 'Golden Country' in *Nineteen Eighty-Four.*

Orwell's hop-picking diary is at https://hoppicking. wordpress.com

5. See my 'The blight of farm suicides', UnHerd 9/1/20 https:// unherd.com/2020/01/whod-be-a-farmer/

> I poshly call it my 'sabbatical'. For the past year, I have been in France studying organic agriculture. (The 'bio' sector here is way ahead of ours.) Before crossing the Channel I was anxious about integration, assimilation, fitting in. Needlessly, as it happens. I talk to my farming neighbours in deep France about exactly the same things as I talk to my neighbours in Herefordshire *profonde*. The weather. The state of the crops/ livestock. The price of wheat . . . The local farmer who's killed himself.
>
> More than one French farmer a day, by some estimates, takes his own life. (And it is almost always 'his', which is part of the problem.) Suicide down on the farm is not an international league table one wants to top: in the UK, the agricultural

suicide rate is a mere farmer per week (still more than twice the national average). But then we have fewer farmers, 138,000 as opposed to France's 450,000.

In France, the topic of agricultural sector suicide is currently on the lips of *tout le monde*, because farmers are part of the national fabric, because of the success of the recent film *Au Nom de la Terre*, directed by Edouard Bergeon, which recounts the life of farmer Pierre Jarjeau, from 1976 to his self-killing in 1996. It is painful watching: all the more painful because it is true. It is the story of Bergeon's father.

The reasons for the grotesque suicide rate in agriculture? They are all there in *Au Nom de la Terre*, and they begin with money.

There is none. In the 1950s (my grandfather's era), a farmer with a small herd of milk cows, say 30 Jerseys, could make a living. In 2020, a farmer with 300 Holsteins will struggle to make a penny/cent profit per pint. Personally, I do not know any farmers who work fewer than 80 hours a week, or for more than the national minimum wage.

Modern farming is a bonfire of the economic sanities, where prices are continuously driven down by supermarkets, who then take the wolf's share of the remaining price of the packet on the shelves. The farmer's conventional answer to the cheapening of food has been to increase production. The Bigging-Up has required pushing machine and man and land ever harder. More hours! More acreage! More technology! (Pierre Jarjeau in *Au Nom de la Terre* invests in a state-of-the-art, automated chicken shed, financed by the sort of easy credit everybody from City suits to EU bureaucrats like to slip farmers.)

The miraculous, eye-wateringly expensive technology goes wrong. Just one piece of kit failing – such as the automated feeders in M Jarjeau's hen shed – can bring the whole precarious Ponzi scheme of your farm's financing crashing down to earth.

Modern farms are kept going only by billions of euros/ pounds of subsidy. A Welsh sheep farmer on the Brecon hills, for example, will get about 80% of his farm income from subsidy. Sounds great. But farm subsidy is a double-edged plough-blade. The basis of subsidy has shifted so often, it is dizzying – farmers have been paid to produce, then not produce, then for the number of head of livestock, then for their acreage – which makes long-term planning impossible. Also, laugh not, but the form-filling of contemporary farming takes hours per day. Online. With rural broadband. I once spent three hours 'pre-notifying' a single pig movement electronically via the eAML2 system. (That 'eAML2' speaks volumes.) It was time wasted. Meaning money lost. A perpetuating circle of poverty. No forms filled = no subsidy. Fill in forms = no farming done.

It's all rules and regulations down on New Macdonald's farm, with its ever-expanding office. The majority of farmers voted for Brexit. Why? You should have tried dealing with the sprouting of tape from Brussels.

More surreptitiously, subsidy strikes at the soul of the farmer. Farmers like to think of themselves as the last heroic individuals, hard people doing a hard job. But the past 50 years have turned us into slaves on subsidy which, to be brutal, is a state handout.

And we are at everybody's beck and call, the kites at the end of the string. My grandfather, who farmed for 50 years, used to call farming 'crisis management'. (Actually, he said rather more earthily, 'It's one bloody thing after another.') The job is by its very nature somewhat tricky, because it is the attempted control of nature. So farmers expect the variable weather, even the plagues such as foot-and-mouth. (Mind you it hurt, the suffering of the animals, the black pyres of their culled bodies.) But my grandfather did not expect his product price to be affected by global markets, City investors speculating in land, inconstant supermarkets paying a pittance and then paying late, Instagrammers influencing what's in and what's out on the national plate, 're-wilding' urban journalists launching a culture war against sheep, or politicians in far-away places making legal ties that bind.

Stressed? You bet. Then there is that dark corner of the mind that knows, just knows, that the more stressed you are, the more likely you are to have an accident, or to become ill. Can't work? Farm goes under. (Farming, by the way, has the worst safety record of any job in the UK, with stress frequently the causal factor. According to the Health and Safety Executive farming accounts for 1% of workers but 22% of all worker fatalities.)

Farmers tend to toil in isolation, as befits 'rugged individualists'. But these days it is entirely possible to spend days in a tractor cab all by oneself, isolated not just from people, but from the very land you work.

Add alienation to the farmer's woes. Today farmers supervise open-roofed units of production instead of husbanding fields. Symbolically and poignantly, in *Au Nom de la Terre*, M Jarjeau suicides by drinking a cocktail of pesticides and herbicides – the very proofs of the new industrialised agriculture.

Farmers don't talk about the money, the isolation, the alienation, because it is not exactly what manly 'rugged individuals' do, is it? And, anyway, farmers don't have time to talk, because time is money . . .

6. My manifesto for ethical carnivorism can be found at https://unherd.com/2020/06/let-them-eat-mutton-a-compassionate-carnivores-manifesto/

BIBLIOGRAPHY

Anon, *The Country Gentleman's Magazine*, vol. I, 1868, vol. II, 1896

Patrick Armstrong, *The English Parson-Naturalist*, Gracewing, 2000

E. J. W. Barber, *Prehistoric Textiles: The Development of Cloth in the Neolithic and Bronze Ages with Special Reference to the Aegean*, Princeton University Press, 1991

B. B., *The Idle Countryman*, Eyre and Spottiswoode, 1943

—*Letters from Compton Deverell*, Eyre and Spottiswoode, 1950

E. C. Chamberlain, *Glossary of West Worcestershire Words*, Trubner & Co., 1882

D. C. Cox, J. R. Edwards, R. C. Hill, Ann J. Kettle, R. Perren, Trevor Rowley and P. A. Stamper, 'Early Agriculture', in *A History of the County of Shropshire: Volume 4, Agriculture*, ed. G. C. Baugh and C.R. Elrington (London, 1989), pp. 20–26. *British History Online* http://www.british-history.ac.uk/vch/salop/vol4/pp20–26

Peter Davis, *The Diary of a Shropshire Farmer*, Amberley, 2011

Jim Downes, *Eastham Remembered*, privately printed, 1978

J. Dunne, A. Chapman, P. Blinkhorn, R. P. Evershed, 'Reconciling Organic Residue Analysis, Faunal, Archaeobotanical and Historical Records: Diet and the Medieval Peasant at West Cotton, Raunds, Northamptonshire', *Journal of Archaeological Science* 107, 2019

George Ewart Evans, *Ask the Fellows Who Cut the Hay*, Faber, 1956

Christine Faulkner, 'Hops and the Hop-pickers of the Midlands', *Folk Life*, 30:1, 1991

R. C. Gaut, *A History of Worcestershire Agriculture and Rural Evolution*, Littlebury, 1939

A. L. J. Gossett, *Shepherds of Britain*, Constable & Co., 1911

Nick Groom, *The Seasons*, Atlantic Books, 2013

H. Rider Haggard, *A Farmer's Year*, Cresset, 1987

—*Rural England*, Longmans, Green & Company, 1902

Stephen J. G. Hall and Juliet Clutton-Brock, *Two Hundred Years of British Farm Livestock*, British Museum, 1989

Emilyn Hathaway, Jeremy Lake, Adam Mindykowski, *The West Midlands Farmsteads and Landscapes Project: Worcestershire*, Worcester County Council, 2012

A. E. Housman, *A Shropshire Lad*, Ballantyne, 1908

N. H. Jones, *Hop-picking – The Nailmakers' Holiday*, privately printed, 1991

Marek Korcynski, Michael Pickering and Emma Robertson, *Rhythms of Labour*, Cambridge University Press, 2013

Alex Langlands, Peter Ginn and Ruth Goodman, *Victorian Farm*, Pavilion, 2008

John Lewis-Stempel, *Young Herriot*, BBC Books, London, 2011

—*Meadowland*, Doubleday, 2015

—*The Running Hare*, Doubleday, 2016

—*The Wood*, Doubleday, 2017

S. J. Looker, *Jefferies' England*, Constable, 1937

Bernard Lowry and Mick Wilks, *The Mercian Maquis*, Logaston, 2002

John McNeillie, *Wigtown Ploughman*, Birlinn, 2012

Malvern Hills District Council, *Tenbury Wells Conservation Area: Appraisal and Management Strategy*, 2008

Richard Muir, *Shell Guide to Reading the Landscape*, Michael Joseph, 1981

National Sheep Association, *British Sheep*, 9th edn, Malvern, 1998

Susan Oosthuizen, 'The Anglo-Saxon Kingdom of Mercia and the Origins and Distribution of Common Fields', *The Agricultural History Review*, vol. 55, no. 2, 2007, pp. 153–80. *JSTOR*, www.jstor.org/stable/40276163

George Orwell (Eric Blair), 'Hop-picking', *New Statesman and Nation*, 17 October 1931

Mark Overton, *Agricultural Revolution in England: The Transformation of the Agrarian Economy 1500–1850*, Cambridge University Press, 1996

F. G. Payne, 'The British Plough: Some Stages in its Development', *The Agricultural History Review*, vol. 5, no. 2, 1957. *JSTOR*, www.jstor.org/stable/40272824

William Pitt (Board of Agriculture), *General View of the Agriculture of the County of Worcester*, Sherwood, Neely & Jones, 1813

William Plomer (ed.), *Kilvert's Diary 1870–1879*, Penguin, 1977

W. T. Pomeroy, *General View of the Agriculture of the County of Worcester*, B. Millan, 1794

Eileen Power, 'The Wool Trade in English Medieval History', 1941, republished in the McMaster University Archive for the History of Economic Thought, http://socserv.mcmaster.ca/econ/ugcm/3ll3/power/WoolTrade.pdf

Francis Pryor, *Farmers in Prehistoric Britain*, Tempus, 1998

Oliver Rackham, *The History of the Countryside*, Dent, 1986

Peter J. Reynolds, *Ancient Farming*, Bloomsbury, 1987

John Stewart Collis, *The Worm Forgives the Plough*, Penguin, 1973

Strategic Stone Study, *A Building Stone Atlas of Worcestershire*, English Heritage, 2012

A. G. Street, *Farmer's Glory*, Faber, 1932

Bridget Sudworth, *Gertrude: Yeoman Farmer's Daughter*, privately printed, 2016

Anne Sverdrup-Thygeson, *Extraordinary Insects*, Mudlark, 2019

J. Thirsk (ed.), *The Agrarian History of England and Wales*, Cambridge University Press, vol. IV, 1967; vol. V, 1985; vol. VI, 1989

—*The English Rural Landscape*, Oxford University Press, 2003

Edward Thomas, *Collected Poems*, Faber, 2004

Flora Thompson, *A Country Calendar*, Oxford University Press, 1984

Robert Trow-Smith, *English Husbandry*, Faber and Faber, 1953

—*Society and The Land*, Cresset Press, 1953

—*Life from the Land: The Growth of Farming in Western Europe*, Longmans, 1956

—*Man The Farmer*, Priory Press, 1973

M. E. Turner, J. V. Beckett and B. Afton, *Farm Production in England 1700–1914*, Oxford University Press, 2001

Martin Wainwright, *Wartime Country Diaries*, Guardian Books, 2007

Philip Walling, *Counting Sheep*, Profile Books, 2015

—*Till the Cows Come Home*, Atlantic, 2018

Lawrence Weaver, *Painter of Pedigree*, Unicorn, 2017

Ralph Whitlock, *A Short History of Farming in Britain*, John Baker Ltd, 1966

Henry Williamson, *The Story of a Norfolk Farm*, Faber, 1941

Tom Williamson, *The Transformation of Rural England: Farming and the Landscape, 1700–1870*, Exeter University Press, 2002

Stephen Woods, *Dartmoor Farm*, Halsgrove, 2003

William Youatt, *Sheep: Their Breeds, Management and Diseases*, Baldwin & Cradock, 1837

WOODSTON: A PLAYLIST

(With thanks to all who made suggestions via Twitter.)

Big Big Train, 'Meadowland', 2017
Blue Oyster Cult, '(Don't Fear) The Reaper', 1976
J. S. Bach, 'Sheep May Safely Graze', 1713
The Band, 'King Harvest (Will Surely Come)', 1969
Samuel Barber, 'Adagio for Strings', 1936
Big Country, 'Harvest Home', 1983
Benjamin Britten (W. Shield), 'The Plough Boy', 1945
George Butterworth, 'The Banks of Green Willow', 1913
—'Loveliest of Trees', from 'A Shropshire Lad', 1911
—(words by A. E. Housman), 'Is My Team Ploughing?', 1911
Henry Castling, 'Our Threepenny Hop', 1901. Parodied in the hop yards as 'Our Lovely Hops' with, according to George Orwell, the last verse sung as:

> Our lousy hops, our lousy hops
> When the measurer he comes round
> Pick 'em up, pick 'em off the ground
> When he starts to measure he don't know when to stop
> Aye, aye, get in her bin and take the fuckin' lot.

Chaos UK, 'Farmyard Boogie', 1988

Frédéric Chopin, 'The Dragonfly', Preludes, Op. 28, no. 11, 1835–9

Matthias Claudius, Johann A. P. Schulz and Jane Montgomery Campbell, 'We Plough The Fields And Scatter', 1861

Claude Debussy, 'Le Petite Berger' ('The Little Shepherd'), 1908

Copper Family, 'Brisk Young Ploughboy', 1988

Bob Dylan, 'Maggie's Farm', 1965

Edward Elgar, String Quartet in E minor, Op. 83, 1919

—Quintet in A minor, Op. 84, 1918

—Cello Concerto in E minor, Op. 85, 1919

Fairport Convention, 'John Barleycorn', *Tipplers' Tales*, 1978

Ivor Gurney (words by Seosamh MacCathmhaoil), 'I Will Go With My Father A-Ploughing', 1921

—'The Fields are Full', 1920

P. J. Harvey and Harry Escott, 'An Acre of Land', 2018

The Housemartins, 'Me and the Farmer', 1987

Incredible String Band, 'Douglas Traherne Harding', 1968

—'Ducks on a Pond', 1968

The King's Singers (trad., arranged Gordon Langford), 'Blow Away The Morning Dew', 2020

Laurie Lewis, 'Who Will Watch the Home Place', 1993

Hubert Parry, 'Jerusalem', 1916

Pink Floyd, 'The Scarecrow', 1967

Henry Purcell, 'When I am Laid in Earth' (Dido's Lament), 1689

—'The Frost Scene', King Arthur, 1691

Camille Saint-Saëns, 'Le Coucou au Fond des Bois' ('The Cuckoo in the Depths of the Wood'), 1886

—'Poules et Coqs' (Hens and Cockerels'), 1886

Saw Doctors, 'Hay Wrap', 1992

Shell Nature Records, British Birds Series, 'Field and Open Countryside', *c*. 1967

Small Faces, 'Song Of A Baker', 1968

Thomas Tallis, 'Spem in Aliuym', 1570

John Tavener, 'The Lamb', 1982

Traditional, 'The Ox Plough Song'; 'The Painful Plough'; 'Speed The Plough'; 'We Are Jolly Good Fellows That Follow The Plough', 'To Be A Farmer's Boy', 'The Hundred Haymakers', 'All Among The Barley', 'The Farmer's Boy' (possibly written by Charles Whitehead, *c*. 1832)

Traffic, 'John Barleycorn Must Die, 1970

Jethro Tull, 'Heavy Horses', 1978

—'Acres Wild', 1978

The Watersons, 'Harvest Song', 1965

—'Country Life', 1975

Ralph Vaughan Williams, 'The Lark Ascending', 1920

—'One Man, Two Men', 1923

—'The People's Land' (film score), 1942

The Who, 'Now I'm A Farmer', 1974

The Wurzels, 'The Combine Harvester', 1976

—'I Am A Cider Drinker', 1976

Robert Wyatt, 'Pigs . . . (In There)', 1999

XTC, 'Love On A Farmboy's Wages', 1983

Neil Young, 'Harvest Moon', 1992

ACKNOWLEDGEMENTS

Woodston has touched the lives of many. I thank Mary Walker, David Spilsbury, Margaret Yarnold, Bridget Sudworth, Pam Pitt, Edna Knott, Sheila Hince, for their sharing of memories, for their time. My cousin Matthew Lambley provided a trove of photographs of our family at Woodston and kindly compiled photographic essays when I was unable to visit (he owns our grandparents' old house in Tenbury Wells). My absolute heartfelt gratitude goes to the historian Stephanie Mocroft, kind beyond measure, scholarly beyond calibration. David Hill was, as well as friend, a goldmine of useful information about farming.

And, once again, I am indebted to my colleagues at Transworld and The Soho Agency: Susanna Wadeson, Julian Alexander, Alex Christofi, Kate Samano, Hayley Barnes, Ella Horne, Isabelle Wilson, Annette Murphy, Martin Myers.

And, of course, Penny, Tris and Freda Lewis-Stempel.

Finally, I send my felicitations to my fellow 'Amoses', Madeleine, Josephine, Richard, Andrew, Alison, Charles, Rob, Benji, Joanne and Julie.

CREDITS

The author has made every effort to obtain permission for the use of copyrighted material.

The extract from John Masefield's 'August 1914' is reprinted by permission of the John Masefield Estate © 1977

'The Visit' by Sir Gabriel Horn is reproduced by permission of the Master of Sidney Sussex College, Cambridge, copyright 1937.

The photographs on pages 105 and 205 appear courtesy of Matthew Lambley.

Illustrations by James Weston-Lewis.

Map by Liane Payne.

draw down CO_2

↓

put it back as stable soil carbon.

↓

only

1/2 carbon there was 12,000 yrs ago.

We have altered our buffer.

→ put it back into the other sink.